Rolls-Royce had been making piston engines for 40 years before its first jet engine, the Welland, powered Britain's first jet aircraft, the Gloster Meteor, in 1944. In the post-war world, the jet engine was soon embraced as the key to the future for both civil and military aircraft. Meeting the demand for steadily larger and more sophisticated jet engines quickly transformed the business of Rolls-Royce. And half a century later, of course, we can look back on a thousand ways in which jet-engine technology has transformed the world at large.

The evolution of today's hugely powerful engines has been an extraordinary story of continuous incremental improvements, with the occasional leap forward to bigger and better things. So, too, with this book. It first appeared in 1955, and has since then been carefully updated and revised though a further four editions, the last in 1996. This latest edition, though, marks a considerable advance. The layout has been extensively re-designed and the text comprehensively rewritten, to take full account of the enormous progress made on jet engines over the past 20 years.

The result builds on all the strengths that have made The Jet Engine a classic of its kind. Remarkable drawings illustrate the complexity of the jet engine in ways that will still appeal to the lay reader. The explanations of the underlying technology have meanwhile lost none of the rigour to be expected from what has been a recommended university text for a whole generation of engineering students. For both audiences, this book remains an inspiring introduction to a challenging subject.

It also takes account of critical trends in engine design over recent years, none more important than those aimed at reducing noise levels and carbon dioxide emissions. Both issues pose problems that will go on demanding innovative engineering solutions for the foreseeable future. This new edition of The Jet Engine is a timely reminder of the astonishing skill with which these and so many other problems have already been addressed in the past.

Sir John Rose Chief Executive, Rolls-Royce plc

contents

section two
define

THIS SECTION, **COMPONENT DEFINITION**, STARTS AT THE FRONT OF THE ENGINE AND FOLLOWS THE AIRFLOW THROUGH TO THE REAR. IT THEN LOOKS AT THE OTHER COMPONENTS AND SYSTEMS THAT NEED TO BE INTEGRATED WITH THE ENGINE.

section three
deliver

THERE ARE GOOD REASONS WHY THE JET ENGINE DELIVERS **IN SERVICE**: THE NATURE OF THE JET ENGINE DESCRIBED IN SECTION ONE; THE ENGINEERING EXCELLENCE OF SECTION TWO; AND THE ABILITIES TO MANUFACTURE, MAINTAIN, AND ADAPT.

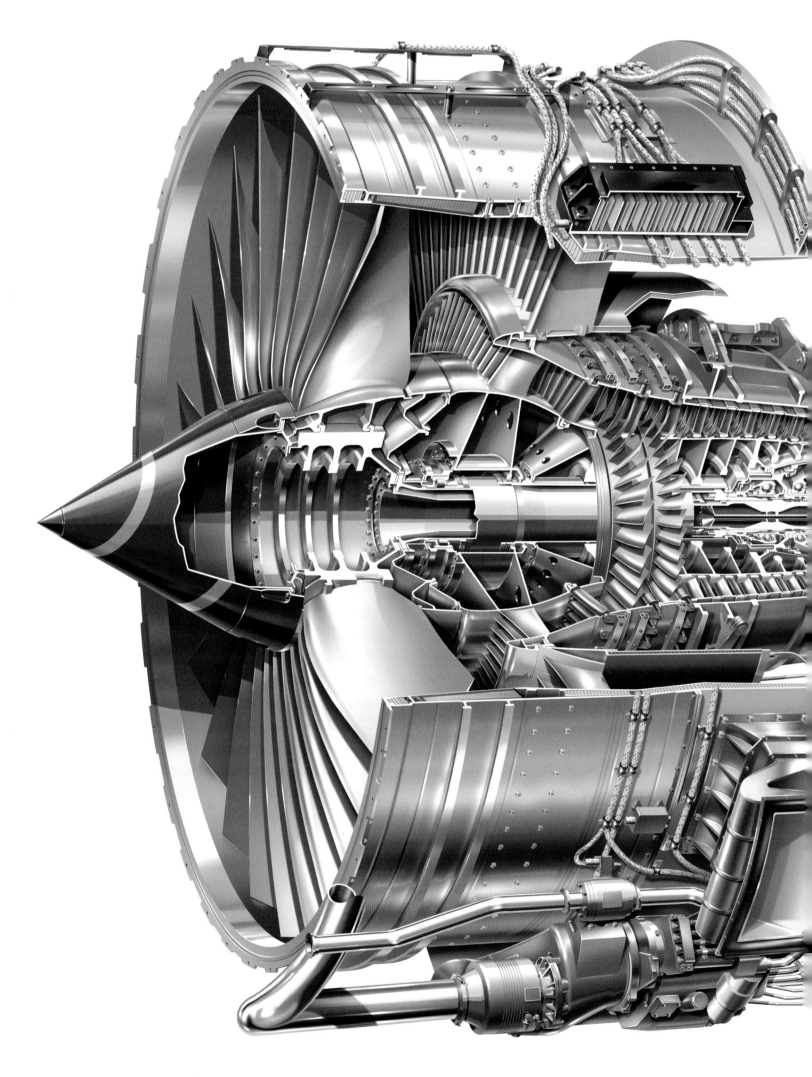

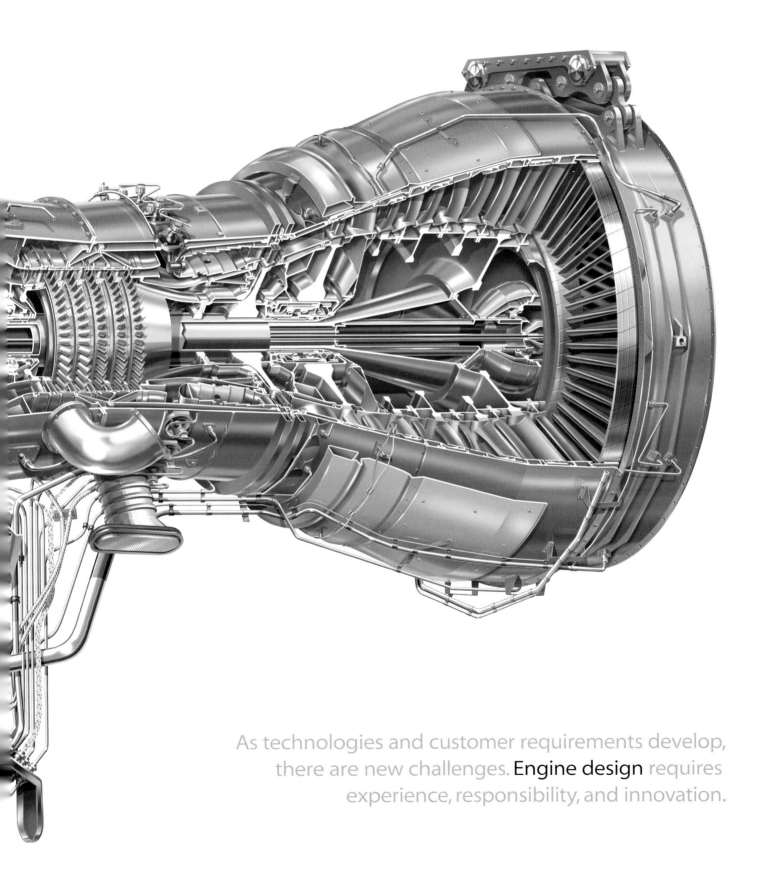

As technologies and customer requirements develop, there are new challenges. **Engine design** requires experience, responsibility, and innovation.

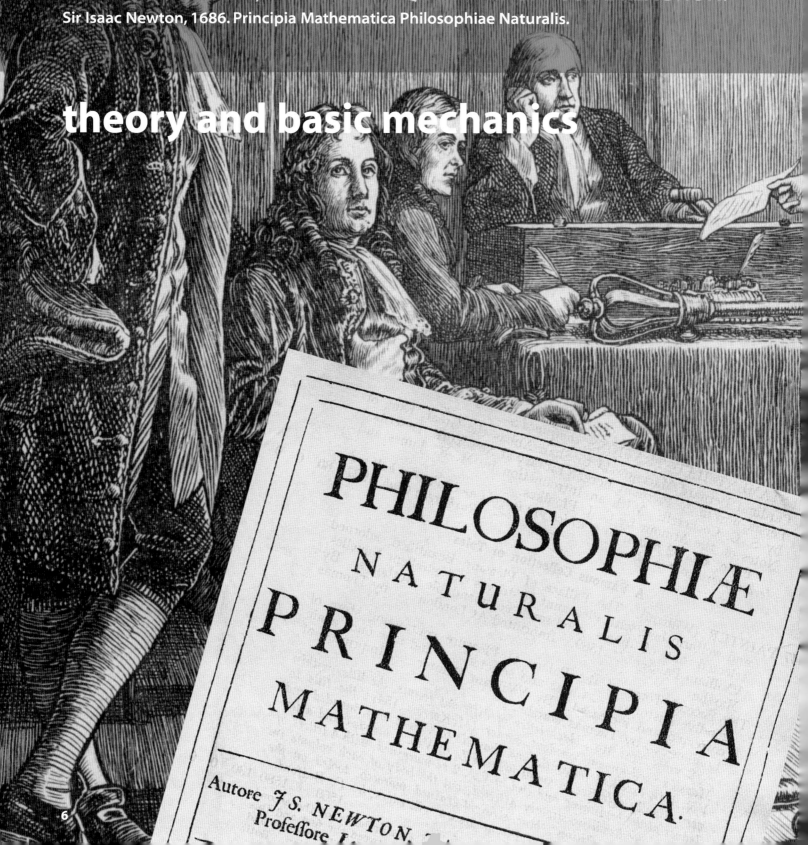

EVERY OBJECT PERSISTS IN ITS STATE OF REST OR UNIFORM MOTION IN A STRAIGHT LINE UNLESS IT IS COMPELLED TO CHANGE THAT STATE BY FORCES IMPRESSED ON IT. FORCE IS EQUAL TO THE CHANGE IN MOMENTUM PER CHANGE IN TIME. (FOR A CONSTANT MASS, FORCE EQUALS MASS TIMES ACCELERATION.) FOR EVERY FORCE ACTING ON A BODY, THERE IS AN EQUAL AND OPPOSITE REACTION.

Sir Isaac Newton, 1686. Principia Mathematica Philosophiae Naturalis.

theory and basic mechanics

❯ How does a jet engine produce useful work,
   where does the energy come from to do it,
   and what is that work used for?

   ❯ How do the internals of a jet engine produce
      work? How does air move through the
      engine, and what happens to it as it does?

      ❯ Why do all large aircraft use jet engines
         instead of piston engines?

         ❯ What are the different types of jet engine,
            and what are their mechanical arrangements?

This chapter provides answers to these initial questions – and, in doing so, inevitably raises more. For example, is it possible to achieve high thrust and high efficiency and a small, light engine, all at the same time?

One of the prerequisite skills of the engineer is to understand the fundamental and contradictory constraints of a jet engine and balance them appropriately for a given design specification. The ideas of balance and constraint are themes that will reappear frequently in the following chapters.

A gas turbine (the type of jet engine described in this book) used on a twin-engined aircraft

Hero's engine or 'aeolipile'. The word aeolipile derives from the Latin 'pila' meaning ball and Aeolus, the Greek god of the winds.

The theory of jet propulsion

Newton's third law of motion states that 'for every force acting on a body, there is an equal and opposite reaction'. The jet engine applies this principle by forcing a fluid, whether liquid or gaseous, in one direction so creating an equal reaction, 'thrust', that moves the engine (and the vehicle it is attached to) in the opposite direction.

The thrust of a jet engine operates on the engine itself – it does not push against the air behind it.

Simple jet engines

A rotating garden sprinkler is a simple, practical example of jet propulsion, rotating in reaction to the jets of water being forced through the nozzles. Hero's engine added

heat to the equation. It was invented around the first century AD, perhaps as a toy, perhaps to open temple doors. Whatever the application, Hero's invention showed how the momentum of steam issuing from a number of jets could impart an equal and opposite reaction to the jets themselves – causing the engine to revolve.

The gas turbine

Most modern jet engines are gas turbines, which are heat engines, and like all heat engines burn fuel to convert their energy into something useful. For a gas turbine, that something useful is a fast moving jet of air propelling an aircraft forward, or powering a turbine driving a load such as an electrical generator, a compressor for a gas pipeline, or a ship's propeller, or water jet.

The gas turbine provides power for many applications: civil and military aircraft, naval and commercial ships, electricity production, gas compression, and oil pumping

Working cycle

The simplest gas turbine, a turbojet, is essentially a tube, open at both ends, with air continuously passing through it. The air enters through the intake, is compressed, mixed with fuel and heated in a combustor, expanded through a turbine, and finally the combustion gases are expelled from a rear nozzle to provide thrust. The turbine drives the compressor via a connecting shaft. This cycle of continuous combustion is known as the Brayton cycle. It defines a varying volume sequence with four distinct stages: compression, combustion, expansion, and exhaust.

The pressure of the gases passing though the engine is always changing. First, pressure goes up in the compressor, it stays almost constant in the combustor (ideally there would be no pressure drop; in fact, it drops marginally), and then the pressure goes down as the combustion gases are expanded through the turbine. The pressure rise in the compressor is usually about twice as much as the pressure drop through the turbine that drives it, so the combustion gases arrive at the back of the engine with spare pressure to accelerate an exhaust jet rearwards.

The relationship between pressure, volume, and temperature

The changes in pressure (and many of the changes in temperature) are caused by changes in the velocity of the air and combustion gases as they pass through the components of the gas turbine engine.

The fundamental laws of compressible flow state that when a gas or fluid is flowing at subsonic speeds through a convergent space (such as a venturi tube), its speed will increase, and its static pressure will decrease. If the gas or fluid flows through a divergent duct, its speed will slow, and its static pressure will increase. This helps to explain the shape of the exhaust and of the passages through the stator and rotor blades of both compressor and turbine.

Boyle's law states that if the temperature of a confined gas is not changed, the pressure will increase in direct relationship to a decrease in volume – and vice versa. Charles's law describes how when a gas under constant pressure is allowed to expand, an increase in temperature will cause an increase in volume – as happens in the combustor of a gas turbine.

In the compressors and turbines, pressure, temperature, and volume are all changing, so Boyle's and Charles's laws need to be applied together as the Universal Gas Law.

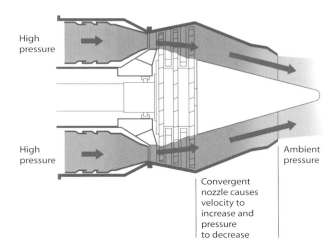

The reduction in flow area causes the gases to speed up and reduce in pressure; this is sometimes called the Venturi effect

High pressure

High pressure

Ambient pressure

Convergent nozzle causes velocity to increase and pressure to decrease

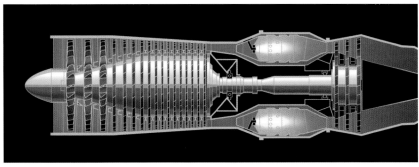

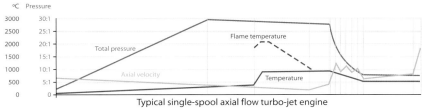

Typical single-spool axial flow turbo-jet engine

The variation of temperature, pressure, and velocity through a simple turbojet

Pressure - volume diagram

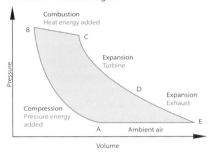

Temperature, pressure, and volume vary through the gas turbine cycle of compression, combustion, and expansion through the turbines and exhaust

Brayton cycle

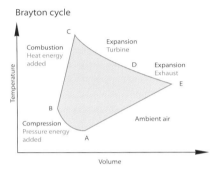

In combination with the reduction in annulus area, the turbine's blades use the same Venturi principle to increase the gas velocity and so the amount of work extracted

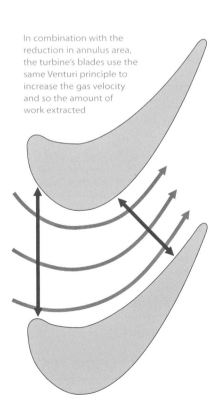

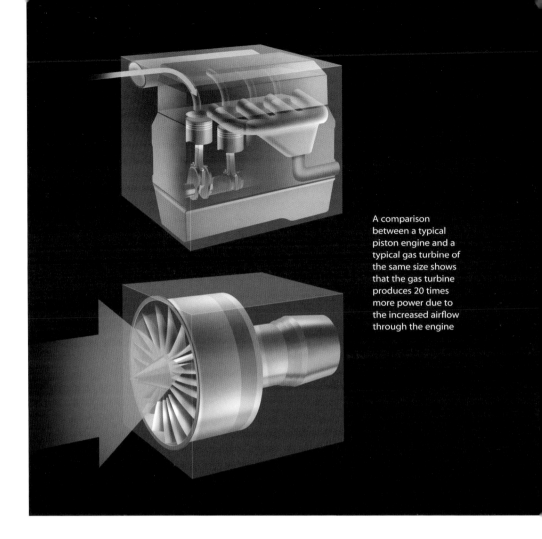

A comparison between a typical piston engine and a typical gas turbine of the same size shows that the gas turbine produces 20 times more power due to the increased airflow through the engine

Producing useful work

The fundamental laws of thermodynamics show that the power required for a given pressure ratio or extracted for a given expansion ratio are directly proportional to the entry temperature. The turbine entry temperature can be five times that of the compressor entry temperature; therefore, the turbine needs a much lower expansion ratio to drive the compressor than the compressor needs to do its work. The difference becomes available to produce thrust when exhausted from the nozzle.

In short, for a simple gas turbine, the hotter the engine is run, the greater the spare pressure and the higher the jet velocity.

The advantages of a gas turbine

Studies (》 288) suggest that the core of a gas turbine can be about twenty times as powerful as the same size piston engine. This is because the continuous cycle and large, open flowpath of a gas turbine can admit 70 times as much air as an equivalently sized piston engine over the same time period. This would suggest that 70 times more fuel could be burnt, leading to 70 times as much energy released in the gas turbine. However, not all the air is used for complete combustion with the fuel. With the assumption that one third of the oxygen in the air passing through a gas turbine is used for combustion, (whereas a piston engine uses nearly all of the oxygen) the energy release rate is about 23 times (70/3) higher than a piston engine of the same size. The ratio of energy release rate varies with size; a comparison of large engines will give different energy release rates from a comparison of small engines.

Being able to move more air through an engine and therefore burn more fuel means that gas turbines can be very powerful for a given size. However, a gas turbine is costly to manufacture because expensive combustor and turbine materials are needed to withstand continuously high temperature. Gas temperatures and pressures can be higher in a piston engine but only at certain points in the cycle; overall, the average temperature in a piston engine is much lower, so the materials used can be cheaper.

Aircraft climbing just after take-off

The gas turbine as an aero engine

For an aero engine, the thrust transmitted to the airframe can be given by the mass flow of air passing through the engine multiplied by the increase in speed of that air.

Air approaches the engine at the flight speed V_{flight} and is ejected faster from the rear nozzle at a speed of V_{jet}. If the mass flow is W, then the thrust F is given by the equation

$$F = W(V_{jet} - V_{flight})$$

This is known as momentum thrust; this equation applies when the nozzle is not choked, and V_{jet}, therefore, is less than Mach one – the speed of sound.

For an unchoked nozzle, there are two ways to increase thrust at given flight speed and altitude. The mass flow W passing through the engine can become larger or V_{jet} can be increased. To increase the mass flow, the engine must have a larger frontal area; it will be bigger, heavier, and produce more drag. On the other hand, a higher V_{jet} makes the engine noisier and increases the fuel consumption needed to obtain a given thrust. The task of the aero engine designer is to obtain a compromise between these two factors.

When the nozzle becomes choked, V_{jet} is fixed at Mach one, and, in order to calculate F, a new term, pressure thrust, is added to the equation

$$F = W(V_{jet} - V_{flight}) + A(p_{exit} - p_{inlet})$$

where A is the jet exit area of the exhaust nozzle, p_{exit} is the static pressure at the nozzle exit, and p_{inlet} the static pressure at engine inlet. With V_{jet} fixed at Mach one, the new term for pressure thrust allows thrust to be increased by raising p_{exit}. This is achieved through a higher total pressure in the jet pipe. Although V_{jet} is fixed at the speed of sound, by running the engine hotter, the speed of sound can be increased, V_{jet} goes up and momentum thrust increases.

The first task of the aero engine is to accelerate the aircraft down the runway. A big engine like the Trent 500 swallows and ejects 1,000kg or one tonne of air every second during take-off. At sea level, one cubic metre of air has a mass of about one kilogram, so the engine is ingesting about 1,000 cubic metres of air every second. If this volume of air were a cylinder the diameter of the intake, stretching out in front of the engine, it would extend for 200 metres – and would be consumed by the engine in one second.

Air is required to provide propulsion for aero engines – the mass of air does not change through the engine, though it does gain energy through the addition of fuel

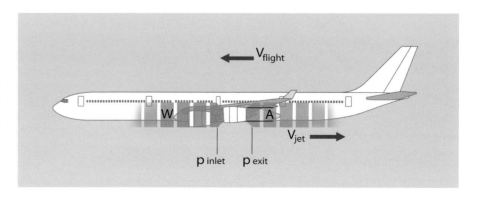

The next task for the engine is to make the aircraft lift off. For example, an Airbus A340-600 aircraft weighs 368 tonnes; each of its four Trent 500 engines produces about twenty-five tonnes of thrust at take-off, giving a total output of 100 tonnes of thrust. Vertical take-off, therefore, is not an option but because the aircraft is going forwards, air passes over the wings and can be turned downwards to create lift. At take-off, a wing gives more than one tonne of lift per square metre – the A340 has 437 square metres of wing, so it can get airborne and climb. The engines do not provide direct lift, but are required to push the aircraft through the air, overcoming the drag of the airframe and the lift-induced drag from the wings.

Flight speed increases until engine thrust equals drag. The aircraft can now cruise with constant lift from the wings. It slowly gains height as fuel is consumed and the aircraft becomes lighter. Then, engine thrust is decreased by reducing fuel flow; the aircraft slows down, descends, and lands. This is a typical cruise profile for a civil airliner.

The turbojet – and its limitations

The first jets to fly were turbojets with a single compressor and turbine. The turbojet is a simple, classic design, and, in only a few years, proved to be a fast, powerful engine. However, the turbojet has now largely been superseded because later developments of the gas turbine have proved more efficient for the majority of air travel.

When an engine has reached a steady running condition, the energy input to the engine from fuel is almost exactly equal to the extra jet kinetic energy output (relative to the engine) and the extra jet thermal energy output. Light and sound energy emission and heat loss across the engine is negligible. About half the energy input goes out as extra jet kinetic energy. This proportion is called the thermal efficiency. A thermal efficiency of 100 per cent would mean that all the energy was being turned into jet kinetic energy with no wasted heat; this is a theoretical ideal, impossible to achieve. Conversely, a fire that does no work has zero thermal efficiency by this definition. Some modern gas turbines can achieve a

thermal efficiency of about 45 per cent. Another measure of performance is propulsive efficiency; this is the work done to propel the aircraft divided by the work done by the engine to accelerate the jet of air.

The part of the fuel energy that goes out as jet kinetic energy will vary with V_{jet}^2 because the jet kinetic energy is given by

$$KE = \tfrac{1}{2}WV_{jet}^2$$

But thrust is given by the equation

$$F = W(V_{jet} - V_{flight}) + A(p_{exit} - p_{inlet})$$

So, thrust will increase in proportion to V_{jet}, but fuel consumption varies with V_{jet}^2.

Therefore, although thrust increases with increasing jet velocity, fuel consumption increases more quickly. This is the tragedy of the turbojet: a high jet velocity, which can be in excess of 1,000 metres per second for simple turbojets, produces high fuel consumption for a given thrust and can be unacceptably noisy.

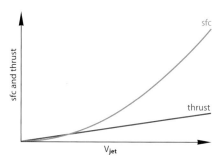

Specific fuel consumption (sfc) increases sharply with V_{jet} compared to the linear increase of thrust

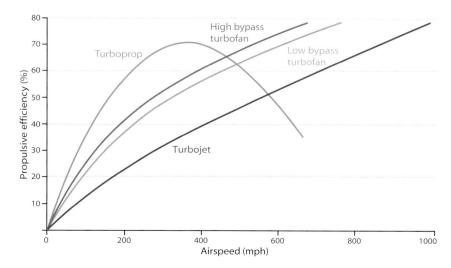

The variation of propulsive efficiency with speed and engine type

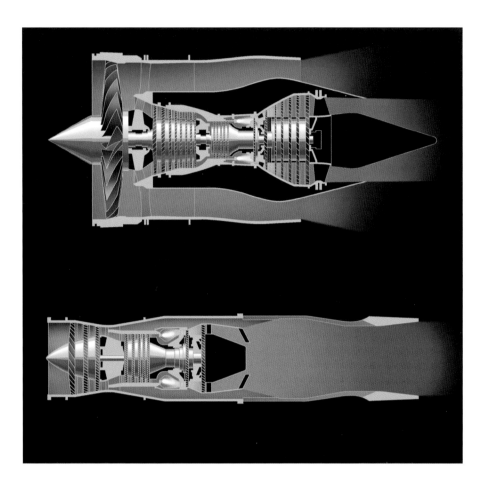

Top: a high bypass ratio
three-shaft civil engine

Bottom: a two-shaft
military engine with
a low bypass ratio
and afterburning

Turbofan types

The core is sometimes called a gas generator because it generates a useful, continuous flow of hot, high-pressure gas at exit from the core turbines. This hot, high-pressure gas can become the single, very high-speed exhaust of a turbojet, or it can be expanded to drive an LP turbine. In a conventional turbofan, the LP turbine is used to drive the fan. The bypass air may then eject from a separate bypass nozzle, or from an integrated nozzle shared with the core flow.

The Trent and the EUROJET EJ200 are both turbofans but are very different in design as they are intended for very different applications. The Eurofighter Typhoon, powered by the EJ200, can fly nearly three times faster then the commercial airliners powered by the Trent (» 75), and so the three-stage EJ200 fan has a higher pressure ratio than the single-stage Trent fan. Coupled with the low bypass ratio this gives the higher jet velocity necessary for higher flight speed.

A low bypass engine with a three-stage fan is the correct choice for the Typhoon because its mission is not always to fly at maximum speed; it must also cruise, loiter, and intercept as a single aircraft system. This contrasts with an interceptor, where a pure turbojet may be the better choice for its typical, high-speed mission. In situations where thrust is more important than noise or fuel consumption, aircraft can use afterburning – burning extra fuel in the exhaust for short periods to gain extra thrust.

Turboshafts and turboprops

Turboshaft and turboprop engines are gas turbine engines where all the useful power output is transmitted by a shaft. Engines that drive an unducted fan or a propeller

The advantages of a turbofan

There are good reasons for an engine to have a high compression pressure ratio and a high turbine entry temperature. However, if all the spare pressure that this generates at the exit of the engine is only used to accelerate the core airflow, the high jet velocity is noisy and does not give the highest possible amount of thrust for a given amount of fuel. The solution – proposed by Frank Whittle (» 26) – is to add an additional low-pressure turbine downstream of the core turbine; this powers a fan to drive additional air outside the core of the engine, through a bypass duct.

The low-pressure turbine, which may consist of several turbine stages joined together, extracts energy from the moving exhaust gases so that, by the time these gases reach the final core nozzle, their pressure and temperature are much lower. As a result, the core jet accelerates to a much more modest velocity, sufficiently greater than the flight speed to create thrust but not so much greater that it creates more noise and uses more fuel.

The low-pressure, or LP, turbine of a Trent 500 extracts 80,000 horsepower from exhaust gases, which it then transmits along a shaft to the large fan at the front of the engine. This fan gives a small pressure rise to a large amount of air, which is then split: some goes through the core of the engine in the same way as a turbojet, while the remainder goes through the bypass duct. Because the fan pressure ratio of the single-stage fan is low, the bypass jet velocity is only slightly greater than the flight velocity.

So, a turbofan engine gets its thrust by accelerating a large mass of air to a modest jet velocity. Since thrust is proportional to V_{jet} but fuel consumption goes with V_{jet}^2, the turbofan gives about twice as much thrust for the same fuel consumption as a turbojet of the same core size. It is also much quieter and so may be used at commercial airports. This could be described as the triumph of the turbofan.

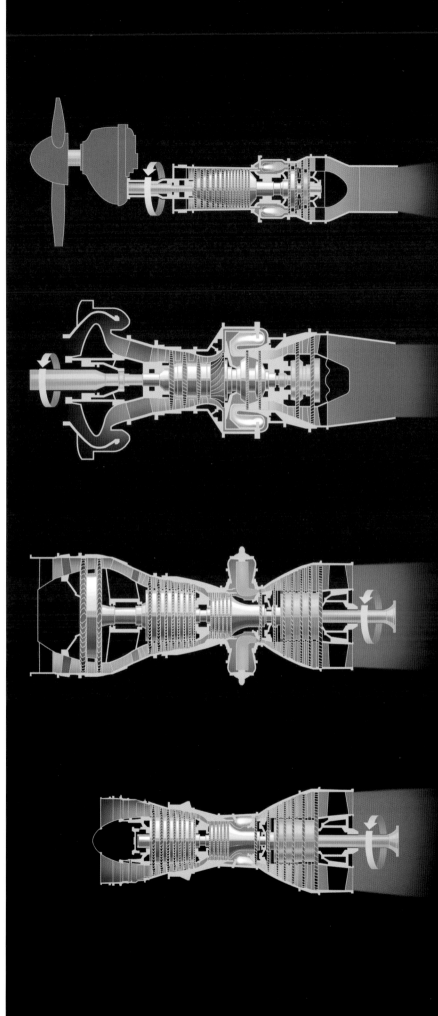

are called turboprops, while the engines that power helicopters are called turboshafts because the helicopter rotor is quite separate from the engine. Turboshafts also drive ships' propellers, generators in power stations, oil pipeline pumps, and natural gas compressors.

A turboprop engine uses the LP turbine to drive a large propeller though a speed reduction gearbox. For a given engine weight, a turboprop, with its large propeller, accelerates more air than a turbofan to a lower velocity, and hence delivers more thrust for a given fuel consumption. Turboprops are lighter than turbofans of the same size because they do not need a nacelle around the propeller. However, the low jet velocity means that as flight speed increases, thrust lapses quickly. This is a factor in preventing the use of turboprops in high-speed applications.

A helicopter turboshaft engine uses LP turbine power to drive a shaft to turn the main rotor. Helicopter rotors are much larger than propeller blades because, without wings to generate lift, a helicopter needs to generate a lot of thrust for lift off.

The Industrial Trent uses LP turbine power to turn a two-stage LP compressor and extracts enough power to drive a 40-50MW external generator or other loads such as a oil pump or a gas pipeline compressor. Marine and industrial engines are similar to the aircraft engines from which they are often derived, but may have heavier components because weight is less important than, for example, low emissions. Marine engines and industrial engines running offshore have special coatings to cope with the salt in sea spray and the sulphur in marine fuel.

An Airbus A340-600 with four Trent 500 engines

Mechanical arrangements

Most gas turbine engines have axial (rather than radial, or centrifugal) compressors and turbines. Axial compressors and turbines consist of sets of rotor blades radiating from rotating discs, interspersed with stationary blades fixed at their outer circumference in the engine casings. In a compressor, the stationary blades are called stators; in a turbine, they are called nozzle guide vanes. The air passing though the compressor rotors and stators is compressed. The task of the compressor is to achieve that compression as efficiently as possible.

Air passes though the open flowpath of an axial compressor at about 150 metres per second, but aviation fuel only burns at a few metres per second. Therefore, prior to combustion, the compressor exit air has to be slowed down before fuel is added through injectors into the combustor flametube.

Once the air/fuel mix is ignited, the flametube provides the necessary protection from the high-speed airflow for flame stability. The rest of the compressor air is fed into the combustor downstream of the stable, primary combustion zone, mixing with the air inside, to give a lower exit temperature profile into the turbine system.

The turbine nozzle guide vanes accelerate and deflect the combustion gases. These high-speed gases move through the turbine rotors pushing them around. In this way, a turbine can generate torque to drive a compressor or fan. The task of a turbine is to do this for the least pressure drop, and to survive for as long as possible at the extreme, continuous temperatures found in the hot end of gas turbine engines.

The pressure built up after the fan and compressor, and left over at turbine exit, accelerates the bypass and core jets through nozzles (or a single, combined nozzle) to obtain thrust. This is transmitted by the engine mounts to the aircraft. If the engine is a turboprop or turboshaft, the last turbine stages drive a load instead of a fan.

The rotating turbine and compressor discs, either individually or joined together into a drum, are attached to the shafts that connect the turbines to the compressors or the power turbine to its load. These shafts are supported by bearings fixed into the engine structure. At the front of the engine, where metal and air temperatures are comparatively cool, ball bearings provide axial location. The rear bearings are typically roller bearings that locate the shafts radially, but allow differential thermal expansion of the shafts and casings in an axial direction.

Multi-shaft layouts

The simplest arrangement of a jet engine has a single compressor, driven via a shaft by a single turbine. In practice, this layout is only used for the smaller turbojets; larger, more complex layouts require a multi-shaft approach.

As the air is compressed on its way towards the combustion chamber, the annulus area of the compressor reduces, and the compressor blades become smaller. In the interests of efficiency, the smaller blades at the rear of the compressor need to rotate at a higher speed than the fan at the front.

This is done by splitting both the compressor and turbine into two: an LP compressor is connected via a shaft to an LP turbine; an HP compressor is connected via a second shaft running outside the LP shaft to a high-pressure (HP) turbine. This two-shaft engine layout is the optimum engine architecture for engines up to 25,000-35,000lb thrust.

Larger turbofans can benefit from three shafts; in this configuration, there is a fan (LP), an intermediate (IP) compressor, and an HP compressor all running on separate shafts connected to respective LP, IP, and HP turbines.

The separation of the fan and first compressor stages allows the shaft speeds and thus fan and blade velocities to be optimised more closely to the ideal operating conditions of each stage.

The three-shaft layout adds a level of mechanical complexity to the overall engine layout but reduces the reliance on variable geometry compressor features. The main benefit is that high thrust can be developed from a shorter, lighter engine than an equivalently rated two-shaft layout.

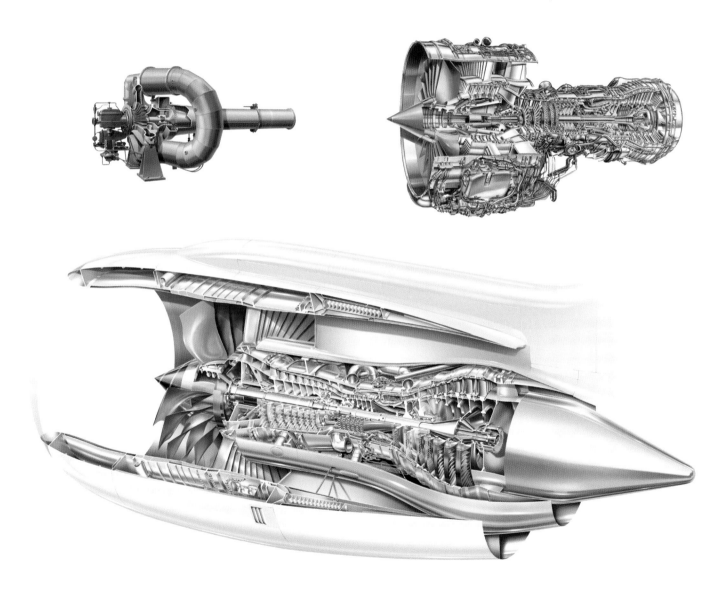

The growth in complexity of shaft arrangements as engine thrust and size increase is shown with the first working gas turbine, Whittle's single-shaft W1, the two-shaft V2500[1] (25,000 to 35,000lbs), and the three-shaft Trent (53,000 to 95,000lbs). Red indicates the HP spool, yellow, the IP spool, and blue, the LP spool.

In theory, there is no difference between theory and practice; experience suggests that in practice, there is.

experience

WHEN FRANK WHITTLE TOLD ERNEST HIVES THAT SIMPLICITY WAS A HALLMARK OF HIS JET ENGINE, THE ROLLS-ROYCE DIRECTOR REPLIED: 'WE'LL SOON DESIGN THE BLOODY SIMPLICITY OUT OF IT!'

OF COURSE, NOT ONCE IN THE HISTORY OF THE JET ENGINE HAS IT BEEN TRULY SIMPLE, NOT IN THEORY, NOT IN MANUFACTURE, NOT IN APPLICATION.

experience

History is usually perceived as a series of distinct and discrete events – indeed the timeline at the end of this chapter shows just such a perspective. Viewed in this way, the history of the jet engine is a rapid procession of achievements, each complete unto itself; collectively, it is a technological progress impressive even by the standards of the twentieth and twenty-first centuries.

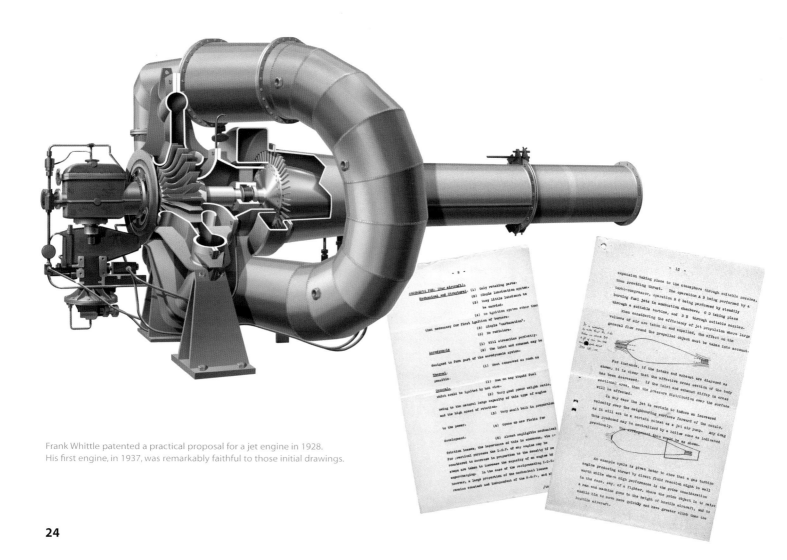

Frank Whittle patented a practical proposal for a jet engine in 1928.
His first engine, in 1937, was remarkably faithful to those initial drawings.

But such a list is only a partial story, and the historical reality is many orders of magnitude more complex. Developing and demonstrating an understanding of what is theoretically and practically possible requires a continuous, laborious, and painstaking search for efficiency and versatility. This search is driven on the one hand by the engineers' intellectual curiosity and passion for excellence, and on the other hand by the customers' desire to use the jet engine in ever more demanding applications. Sometimes, the customer pushes the engineer; on other occasions, the engineer surprises the customer with a new view of what is possible.

The Whittle WU turbojet undergoing testing

Dr A. A. Griffith, his thinking still influences the design of today's jet engines

The early days

In 1903, Orville and Wilbur Wright achieved sustained, controlled, powered flight at Kitty Hawk, North Carolina. Their craft, the 'Flyer', was powered by a 12hp piston engine. The flight lasted some twelve seconds and covered 120 feet; the speed, admittedly against a strong headwind, was barely that of a brisk jog. Twenty-eight years later, a Rolls-Royce R engine, capable of 2,530hp, powered a Supermarine S6B to a new world record of 407.5mph. This was rapid and impressive progress, spurred on initially by amateur enthusiasm, then national pride, and as World War II loomed, national security. But engineers knew there were both practical and theoretical limits to the speeds possible using a propeller and piston engine. Long before the success of the R engine, the search had already begun for an alternative.

In 1922, Maxime Guillaume patented his idea for an axial jet engine – but it remained no more than an idea.

Dr A. A. Griffith was a mathematician and aerodynamicist extraordinaire who worked at the Royal Aircraft Establishment. In 1926, he published an analysis of an axial turbine that led to a rotating test rig of an axial turbine and compressor; this was followed, in 1929, by a proposal for a turboprop – a design so sophisticated it was at least half a century in front of manufacturing capability.

In 1935, Hans von Ohain, a physicist at Göttingen University, proposed a turbojet with both an axial and centrifugal compressor. Supported by the aircraft manufacturer, Heinkel, his was the first jet engine to fly, in 1939.

But it was Frank Whittle, of the Royal Air Force, who patented the first practical proposal for a turbojet in 1928, a patent that became widely available and studied. Whittle was a remarkable aviator and engineer, and his invention, in 1937, was the first turbojet in the world to run, at a not all together controlled 8,000rpm.

Rolls-Royce, aware of these developments, recruited Griffith in 1939 and set him up in the luxurious company guesthouse to 'think' – this thinking, over several years, informed many later designs. Meanwhile, Rolls-Royce was also supporting Whittle with rig testing and by making components such as turbine blades and gearcases, at no cost to Whittle's company. In 1943, Rolls-Royce took over development of Whittle's W.2B engine, then still very much at an experimental stage. Just over a year later, the jet engine was in squadron service. The Gloster Meteor, powered by the Rolls-Royce Welland turbojet, quickly became part of the battle against the V-1 flying bomb. It was the only allied jet aircraft to see action in World War II.

To take a completely new type of engine from concept to combat in sixteen years was remarkable, especially at that point in political and industrial history. Governments recognised that the potential speed of the jet engine could bring military advantage, but were necessarily reluctant to divert too many resources from other areas of the war effort. And, compared to a conventional engine, making a jet engine was a formidable challenge. Compression and combustion occur intermittently in a piston engine but continuously in a jet – and at higher average temperatures, pressures, and speeds: the existing technologies could not cope. The compressors were too inefficient, despite the experience of Rolls-Royce with supercharging the R engine. Making turbine blades that could operate continuously while rotating at red-hot temperatures was a new challenge.

The Welland turbojet on an outdoor testbed

Most difficult of all, on the early engines, was the combustor, which needed to burn fuel at much higher rates than previously attempted, in the middle of an airflow so fast it would extinguish any flame.

Whittle had hoped jet engine design would be an 'exact science'. In those early days, there was a large element of trial and error.

Nevertheless, by the end of World War II, many countries were manufacturing jet engines. One of the early success stories was the Rolls-Royce Nene, which first ran in October 1944 producing 5,000lb thrust; it was later manufactured in Canada, the USA, France, and Russia – it was still being made in China a quarter of a century later.

Pressure, temperature, and efficiency

Throughout the history of the jet engine, engineers have sought to improve its efficiency. Naturally, many factors are involved but three key considerations are the pressure rise achieved by the compressor, the temperature of the gases as they enter the turbine, and combustor efficiency.

Compressors in the 1940s struggled to achieve a 5 to 1 pressure rise; in 2005, the compression system on the Trent 900 had a ratio of 42 to 1. And the turbine entry temperature has risen from 1,000ºC in the 1940s to around 1,700ºC in the twenty-first century. In the 1950s, the early turbojets had a specific fuel consumption above 1.0; specific fuel consumption, or sfc, is calculated as kilograms of fuel used per hour per Newton of thrust. Today, the Trent 800 has a cruise sfc of 0.56 – a 50 per cent improvement.

Obviously, as efficiency and power increase, the range of possible uses for the jet engine also grows.

The Gloster Meteor, powered by the Rolls-Royce Welland, was the only allied jet aircraft to see action in World War II

Designing for civil and military aircraft

The first applications for the jet engine were military aircraft, and the first requirement was speed. However, the post-war years soon saw a demand for passenger aircraft, especially in North America where companies like General Electric and Pratt & Whitney came to dominate the jet engine market. Initially, there was considerable overlap between civil and military requirements, and the same engine could be used in very different applications. The Rolls-Royce Dart, an early, simple, and very successful turboprop, was originally designed for use in an RAF trainer; it in fact powered, among other aircraft, the Vickers Viscount, the world's first production jet-powered airliner. The Rolls-Royce Avon became the benchmark engine in the 1950s for both civil aircraft such as the Comet and Caravelle and many military

aircraft, including the Hunter and Canberra. Notably, it powered the English Electric Lightning, Britain's first supersonic fighter.

The Avon was the first Rolls-Royce production engine to feature cooled high-pressure turbine blades; it was also the first Rolls-Royce engine with an axial compressor – an indication of how difficult it was to design and manufacture an engine based on Griffith's ideas rather than the centrifugal compressor used by Whittle. The effort of developing the axial engine was worth it, though, because of the extra thrust achievable for a given engine diameter.

The technological advances of the Avon paved the way for the Rolls-Royce Conway. With almost twice the thrust and pressure ratio of the Avon, it notched up a notable double first: it was the world's first bypass

engine, and the first to use titanium blades. The Conway powered both the Handley Page Victor bomber – and also the new passenger aircraft like the Douglas DC8 and Boeing 707.

It was not until the late 1950s that Rolls-Royce designed an engine specifically for civil use, the Spey. Even here, a military version was later developed, the RB163 – but this did mark the divergence in requirements. Passenger aircraft required power and economy: attack aircraft needed speed and special performance characteristics at very high and low altitudes.

This is not to say that passengers did not want speed. The popularity of Concorde proved that. The Olympus engine weighed seven times as much as Whittle's first engine, but achieved 25 times the thrust at three

The English Electric Canberra, Britain's first jet bomber, made its first flight on 13 May 1949, powered by two Rolls-Royce Avon RA3 engines. It entered into service in 1951.

While most manufacturing techniques have seen dramatic changes over the last thirty years, the final assembly of an aero jet engine is still very much a highly skilled, hand-built operation

times the speed – and with lower specific fuel consumption. Concorde entered service in 1976 with Air France and British Airways. It flew at twice the speed of sound for three or four hours, every day for 27 years. Compared to that, the average fighter aircraft leads a quiet and pampered life.

But the real trend for passenger transportation was not to go faster, but bigger. Bigger, quieter, cleaner, easier to maintain, cheaper to run.

Revolutionary wide-bodied aircraft like the Lockheed Tristar and Boeing 747 demanded a new generation of large turbofans. The RB211 was one of the first of those turbofans.

It was also the first three-shaft high bypass turbofan, and the first engine to have hollow, titanium fan blades.

At this time, military engines were following some very different paths, one of the most exciting of which was vectored thrust.

The military had always wanted an aircraft with the manoeuvrability of a helicopter and the speed of a jet fighter. Rolls-Royce demonstrated the feasibility of this in 1954 with the Flying Bedstead, otherwise known as the Rolls-Royce Thrust Measuring Rig. From then on, progress in this highly complex field of aviation was phenomenal.

The Harrier, powered by the Pegasus, made aviation history when it entered service with the RAF in 1969 as the world's first front-line, V/STOL (vertical/short take-off and landing) jet aircraft.

Vectored thrust is also a feature of the new Joint Strike Fighter. This, like the Eurofighter Typhoon and other modern military applications, is a multi-role aircraft and as such needs the traditional properties of a turbojet with the versatility and economy of a turbofan. The modern military turbofans, therefore, are very different in design from the latest civil turbofans such as the Trent family.

The Lockheed TriStar was one of the new generation of wide-bodied aircraft. It was powered by three Rolls-Royce RB211-22B engines, each giving 42,000lb thrust.

Civil and military aero engines

Increase in power outputs over time
civil **and** defence

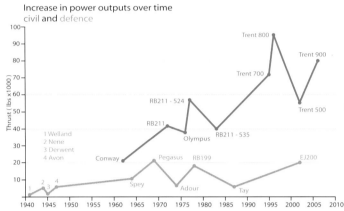

The increase in turbine entry temperature

Increase in turbine entry temperatures over time

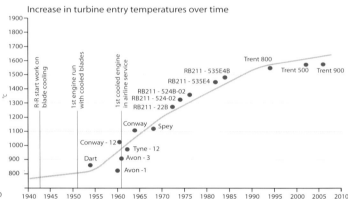

Industrial and marine engines

Increase in power outputs over time
marine **and** energy

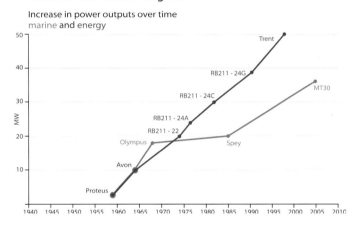

Top left: civil and military aero engines growth in thrust since 1940

Top right: the increase in turbine entry temperature demonstrates improvements in material properties and cooling technology

Left: industrial and marine engines have seen significant and sustained growth since 1940

Another, often unsung, contribution to making efficient engines more efficiently has been the computer. Computer-aided design and computer-aided manufacture, pioneered by Dr. Patrick J. Hanratty in the late 1950s, have transformed the engineering and manufacturing processes. It was originally thought that CAD/CAM would save time and, while this is probably true, its real benefits are more fundamental. Computer-aided design allows the engineer to model and test a design many times over before committing it to metal. Computer-aided manufacture, with computer-controlled tooling, can achieve a precision and consistency that was impossible by hand.

Some of today's engineering practices are only possible because of the immense computational power now available to us. Finite element analysis (FEA) models can be used to analyse the stresses on a material or component. Computational fluid dynamics (CFD) is used to predict and simulate the flow of the gases through the engine.

Together, the silicon chip and the titanium alloy, sprinkled with human intelligence, can take much of the credit for the efficiency of the modern aero engine. Assuming similar thrust, today's engine is half the weight and burns half the fuel compared to a 1950s design. Furthermore, instead of lasting a few hundred hours between overhaul, an RB211-535E4, in 2000, set a world record of 42,000 hours on wing.

Some of today's engineering practices are only possible because of the immense computational power now available

Agents for change: silicon and titanium

Griffith and Whittle, in their different ways, demonstrated that the engineering and manufacturing capability required to make a working engine does not always match the theoretical understanding. One of the challenges throughout the history of the jet engine has been to narrow that gap.

New materials help. Titanium alloys are light in weight and can resist high temperatures; unfortunately, they are also expensive and sensitive to abrasion. However, the use of titanium in components such as discs and blades has transformed jet engine design. Other materials have had a similar impact. For example, ceramics are now used in combustion chambers and on turbines for their mix of low weight and heat resistance.

Early three-shaft engine
on development testing

The global impact of the jet engine

Marine

1960 Brave class
Proteus 3MW

1968 Type 14
Olympus 18.7MW

1959 Power generation
Proteus 3MW

Energy

1940	1945	1950	1955	1960	1965	1970

1950 Viscount
Dart 1,547hp

1960 707
Conway 20,600lbs

1972 Tristar
RB211 42,000lbs

Civil

1945 Meteor
Welland 1,700lbs

1951 Canberra
Avon101 6,500lbs

1960 Lightning
Avon211 14,430lbs

1969 Harrier
Pegasus 21,500lbs

Defence

On land and sea

In 1953, the first Rolls-Royce gas turbines for marine propulsion went to sea, and over the next 12 years Rolls-Royce pioneered the industrial use of aero-derivatives. Here, a jet engine, normally burning natural gas and feeding a free power turbine, drives an electrical generator, a compressor for gas pipelines, or a pump for oil extraction.

Marine and industrial applications obviously have some very different requirements from an aircraft. However, the small size, lightness, and cyclic capability of the aero-derivative turboshaft – all characteristics of the aero jet engine – have been fundamental to its success.

For warships, the aero-derivative turboshaft is now the engine of choice, as space and weight are at a premium. This is also true on offshore platforms and in remote locations, where ease of transportation is also a benefit. For electrical power generation, its fast start-up time and cyclic capability allow use for backup or peaking. Industrial and marine engines operate in harsh, often corrosive, environments, and, for pumping especially, may have to run continuously for days at a time.

The first industrial application of a Rolls-Royce aero engine was to provide backup electrical power. This was the Proteus in 1959 at Princetown, Devon. The Olympus followed three years later, in 1962. In 1964, the Avon

turbojet was adapted for both compressing gas and generating electricity. In 1977, the Industrial Avon set a record of 44,562 hours on gas pumping duty before overhaul; in 2004, the Avon fleet passed 55 million hours. Power generation and gas compression remain common industrial applications today, notably in North America where Rolls-Royce engines power several major oil and gas pipelines.

In 1968, Proteus engines were adapted for use on hovercraft, notably the SRN4 which ferried cars and passengers across the English Channel at speeds up to 65 knots. More conventionally, the same year, the Olympus was adapted for marine use. Over the next decade, it was installed on warships such as HMS Exmouth, the first large

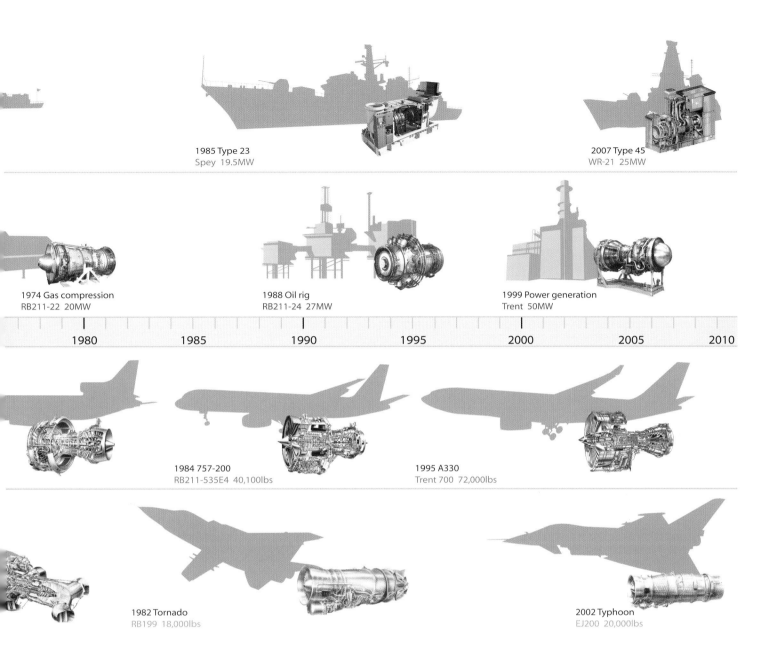

1985 Type 23
Spey 19.5MW

2007 Type 45
WR-21 25MW

1974 Gas compression
RB211-22 20MW

1988 Oil rig
RB211-24 27MW

1999 Power generation
Trent 50MW

1980   1985   1990   1995   2000   2005   2010

1984 757-200
RB211-535E4 40,100lbs

1995 A330
Trent 700 72,000lbs

1982 Tornado
RB199 18,000lbs

2002 Typhoon
EJ200 20,000lbs

warship to be powered entirely by aero-derived gas turbines; the Olympus now powers most major Royal Navy warships including the three aircraft carriers: HMS Invincible, Ark Royal, and Illustrious.

The location of many power generation engines, often near to centres of habitation, requires ultra low emissions of nitrogen oxides and carbon monoxide. Modern industrial engines, such as the Industrial Trent and RB211, are therefore very clean engines, with combustion features not found on today's aero engines. They can use multiple combustor zones or water injection to ensure optimum control of flame temperatures.

Global impact

The jet engine has changed the way wars are fought, the way power is generated, and, with cheap and widely available air travel, it has changed the lives of millions. In 1945, a one-way flight across the Atlantic took fourteen hours. In 1952, the cheapest return flight from London to New York cost more than three months average earnings. In 2003, it cost only four days average earnings and each flight took only eight hours. It is not surprising, therefore, that around two billion aircraft tickets are sold every year. The jet engine has changed the way people travel, and think about travel; arguably, it has altered everyone's perception of the world.

Arguably again, there is a risk that the jet engine might change the world itself.

These environmental concerns are major influences on current engine design – and will continue to be for the foreseeable future.

A continuum of development

The history of the jet engine is an incremental one, continually developing ideas and technologies, building on what is possible at any moment to create a collective body of learning and understanding, which will be continually drawn from and added to. This will be as true tomorrow as it was yesterday and is today.

Past performance dictates future development.

design and development

DESIGN IS BOTH SCIENTIFIC AND ARTISTIC, BOTH PRECISE
AND IMPRECISE; THE DESIGN OF A JET ENGINE IS INVARIABLY
A COMPROMISE OF CONFLICTING REQUIREMENTS.
DEVELOPMENT PROVES THE DESIGN OF AN ENGINE BY APPLYING
EXPERIENCE, INTELLECT, AND THE GRAVEST OF PHYSICAL ABUSE.

design and development

Design: converting requirements into products

All products are a response to a need; they perform a function and have a customer. The design process satisfies some basic human needs by providing artefacts to sustain life; it extends the boundaries of human knowledge, gives pleasure – and provides a great sense of satisfaction to the practitioners.

For most products, the input to the design process is a customer need; the output is a definition of an optimum component, system, or process. The design process consists of two major elements: requirement definition and design definition. These two elements interact with each other; more often than not, both the definitions are iterative processes.

The design process

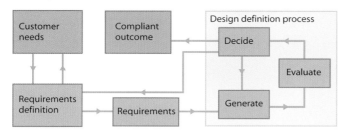

38

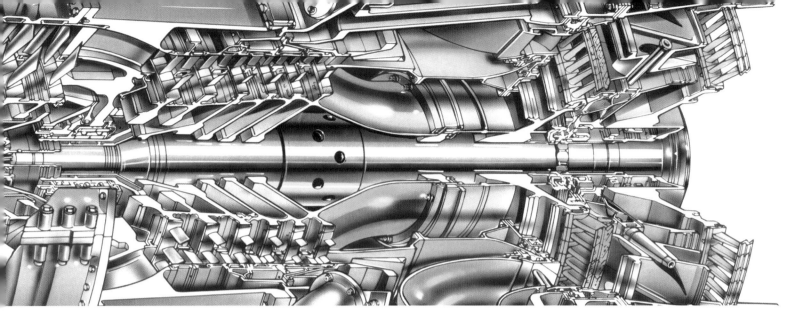

Development: proving a product meets its requirements

The overall aim of the gas turbine product development process, also known as the validation or experimental process, is to achieve certification of the product in line with the regulatory conditions for safe in-service operation and to introduce changes to address

> safety
> efficiency
> performance
> reliability
> operational concerns.

A vacuum cleaner, a Formula 1 racing car, an artificial hip, and a hand-held navigation system: four examples of effective design

Design
Requirements definition

Requirements definition is all about understanding customer needs and translating them into a coherent set of requirements as an input to the design definition process. Requirements should be fully defined before starting the design, and they should not dictate what the solution should be.

Who are our customers?

Customers come in various guises and it's important to know them and their needs. The customer may well express needs in a form that requires interpretation for that information to be useable in the design process: for example, passengers want a quiet aircraft interior; this requires a definition of 'quiet', a sound level in a unit of noise measurement. The certifying authorities also define requirements that the engines have to meet, before and in addition to the customers' requirements. Often, the engine manufacturers demonstrate by test or analysis that these requirements are met.

Every product has a function – something it must do. The function of a gas turbine is to provide thrust or shaft power to drive a load; the components within the engine may have other functions: to heat, to manipulate, to cool, or move, to give just a few examples. Similarly, every product has characteristics, or 'attributes' such as weight, size, unit cost, operating cost, life, aesthetic appeal, and environmental impact.

The customers whose requirements have to be met in the design process

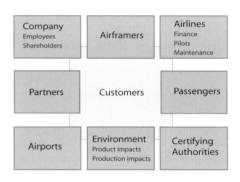

Some attributes are directly related to a product's functions. For example, a turbofan engine has thrust, reverse thrust, and electrical power as function related attributes. Other attributes are associated with the product's interfaces or are created as a by-product of the prime function. For instance, the noise generated by a turbofan is a by-product of the operation. Desired or acceptable limits for attributes are often specified in the customers' requirements.

The typical specification for a turbofan engine includes

> ❭ safety

> ❭ thrust at a number of flight conditions

> ❭ fuel consumption

> ❭ reliability and availability

> ❭ installation compatibility

> ❭ structural integrity

> ❭ unit cost

> ❭ operating cost

> ❭ weight

> ❭ size

> ❭ risk

> ❭ timescale

> ❭ noise

> ❭ emissions.

It is important to know how the customer ranks and prioritises the requirements. Concorde, for example, placed a higher value on speed than on noise or fuel consumption.

Requirements are often written as target values while constraints are defined by a 'not-to-exceed' value. Attributes, therefore, can also have a target value and a not-to-exceed value. Aircraft range depends on weight and fuel consumption so these are target values for the designers and the development team to meet; a pylon has a not-to-exceed value, a structural limit controlling the maximum weight of the engine; the fuel tanks have a finite size that provides a not-to-exceed figure for fuel consumption in order to meet the target range.

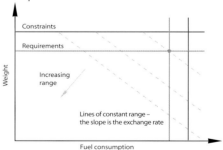

Requirements trade of weight and fuel consumption

Design definition process

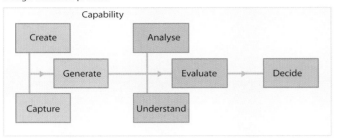

Capability

Create · Generate · Capture · Analyse · Evaluate · Understand · Decide

Design definition process

Three activities form the basis of the design definition process: generating ideas and solutions, evaluating those ideas, and deciding upon the optimum solution.

'Capability' supports the whole design process and can be defined as all that is necessary to achieve the desired results: the people, the technology, the resources, the information, the processes, the facilities. Often, many of these capabilities do not exist when starting out on a design. The things that do not exist obviously add risk to the design and these risks need careful consideration and mitigation as the design develops.

The design process is highly iterative – as ideas are evaluated and improved upon, the design space tightens and the process becomes more like a spiral aimed at the target attributes. Often concepts will be developed to map out the design space and show the strength of the gearing between different attributes. For example, achieving a required thrust can be met with a large range of gas turbine styles: some will be heavier; some, more fuel efficient; others, quieter. The process becomes one of selecting the optimum solution once the primary function target has been met.

Generating ideas and creating concepts requires a mixture of time, information, imagination, knowledge, and experience. Many techniques are available to assist with idea generation and problem solving; normally, people in groups achieve more than individuals in isolation – the interactions and suggestions lead to debate and spark other ideas.

Evaluation is the process of determining how the product will perform when measured against all of the relevant attributes. The simplest form of evaluation is purely comparative: each attribute is scored based on judgement or experience. Early in the design process, where many concepts exist, the evaluation needs to be quick to begin the journey along the design spiral.

Sometimes, where risk exists or the concept is novel, the only way to evaluate an idea is to manufacture prototypes and test them. Clearly, this is expensive and time-consuming, and so great effort is spent on developing analytical computer processes as an alternative way of evaluating the behaviour of different concepts.

The final stage in the process is the decision. This requires both knowledge of the customer and experience of product and process; usually, the more important attributes will need a weighting applied, but it is often impossible to determine a clear winner from an evaluation. The engineer then has to apply judgement and experience to select the right concept.

From design to development

When designing a jet engine, the judgement and experience used by the engineering teams is channelled and guided by a formal review process that covers not only the design and development functions but also the entire life-cycle of the engine – and is applied to the complete product, subsystems, and individual components.

This review process can be broken into seven stages: innovation and opportunity selection, preliminary concept design, full concept design, product realisation (or development), production, continuing service support, and disposal. Naturally, many tasks in the design and development phases overlap; however, formal gates ensure that progression between key points in the process only occur after peer analysis and review.

Design is necessarily concerned with form, material, and function; alongside these, many other factors that have an impact on the final design are also considered: technology requirements, manufacturing capability, supply chain capability, and cost, to name only a few. In the preliminary phase, assumptions about these can be made, which have to be defined in the full concept design.

After many iterations, the design is established. It now has to be validated.

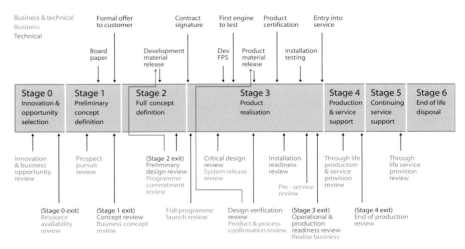

The design and development phases and their position in the product life-cycle

Development

The development objectives are to

› validate that the new product will function according to its specification

› verify that the new product is compliant with all its requirements

› standardise modifications to previously certified products, proving both compliance with certification requirements and that design changes actually address the problem.

Utilisation of core technology across various business sectors allows the gas turbine manufacturer to apply common product development. Differences occur in the certification process as the product requirements differ depending on the application.

The experimental process

In theory, product development begins after the preliminary and full concept definition. In reality, the business pressures to design, develop and introduce new products and modifications to market in ever quicker timescales mean that certain activities within the experimental process, also known as validation, run in parallel to the design process – for example, certification clearance strategies, pre-planning of the engine/rig development programme, and participating in sub-system risk reduction.

Product development then continues for the whole life-cycle of the engine, and so the experimental process is also applied during and after the production phase after engine type certification. During the product's in-service operation, unforeseen technical problems may arise; the regulations may be amended, or customers' requirements may change: modifications to the certified product may be required to accommodate these occurrences.

The experimental process starts with one of the following:

› a new or derivative engine requirement from an integrated system supplier (an airframe or platform manufacturer)

› an internally perceived engine requirement or modification

› a change to a certified engine standard

› a significant service problem on an existing engine type

› a change to the regulations (noise or emissions, for example).

It is then complete when there are fully verified and validated:

› data

› engine components, subsystems, and accessories

› methods of manufacture, engine assembly, test, and strip

› whole engines

› design changes

› documentation.

Full gas turbine certification and in-service modification programmes

Where the requirement is from an integrated system supplier for a new or derivative engine, a full certification programme is usually required. The full suite of experimental activities must be completed to ensure full product development and certification. It is this process, with its emphasis on ensuring validation and certification, that lies at the heart of product development – not any individual test, however spectacular or dramatic.
The development process is structured in five phases: the planning phase, in which the strategy and certification process is planned and agreed with the certifying authority; the programme implementation phase when the experiments are designed and the instrumentation specified; manufacture and assembly of the development engine; performing the required tests; and, finally, analysis of the test data.

Risk reduction

The reduction of risk must be considered from the onset of the product design. This includes the risks associated with the experimental process as applied to all elements of the product definition:

› product physical and functional design

› assembly methods

› usage instructions.

The development engineer liaises with design, manufacturing, and assembly departments and uses whole engine build, test, and flight experience of similar designs to appreciate – early in the programme – the risk areas within the definition and construct the test programmes accordingly.

Compliance strategies

For an identified major change, a compliance strategy must be produced to identify which tasks (and any interdependencies between them) need to show compliance with the internal requirements and those of the airworthiness authority. The identified tasks will come from a combination of the airworthiness authorities' prescribed certification tests, strategy reports, and the identified risks for the element of the product under consideration.

Execution and reporting of the experiments

The identification of the risk and associated mitigation action combined with the compliance strategy will define the experiments' requirements. Experiments compiled to satisfy the product development validation requirements are performed on experimental test vehicles (engines). Each test vehicle may address more than one experiment through bench or flight manoeuvres up to and including limitations and safety requirements. The testing will include experiments to understand and fix problems and to check project and airframe or platform requirements such as performance and noise.

There is also a requirement to report on the outcome of a vehicle test in all cases and to report on its component strip condition when that data is required for the whole engine and not just an individual experiment.

Verification of the production assembly, strip, and test methods

During the experimental process, the production assembly, strip, and test methods will have been verified. To ensure that the new production build and test factory correctly implement the engineering instructions, a technical risk assessment process is employed, identifying the hazard areas and ensuring they are addressed.

The engine development plan

Development programme definition is an iterative process that starts with a set of technical and programme requirements and ultimately ends up with a costed programme definition known as the Engine Development Plan (EDP) that has taken into consideration:

> internal project and external customer validation requirements

> airworthiness authority or classification society requirements

> build and test facility capacity and capability

> build, test, and strip lead times

> engine parts required to deliver evidence

> non-engine hardware and build tooling required to deliver the programme.

The EDP is produced by the development organisation and consists of the following elements:

> a time-based plan of all test vehicle test slots required and when the major certification and production hardware will be available

> a listing of all the assets required to execute the programme according to the plan including engine hardware, build tooling, and slave test equipment

> the manpower resources required to execute the programme to plan

> a risk management plan for the programme

> the budget to deliver the above.

The certification and standardisation process
Civil aviation requirements

The engine certification process for civil engines begins with the identification of a requirement to certify or validate a new civil engine, or amend approved operating conditions of an existing engine.

The certification requirements are to show that the engine has a suitable level of strength, reliability, and safety so that hazardous in-service events are minimised. This is achieved by demonstrating that the engine meets the benchmarks as defined by the airworthiness authority.

The civil certification process

Planning and consultation phase:

> New engine type or extensive change to existing engine type requirement agreed.

> The Chief Engineer, in consultation with airworthiness department, defines the likely certification requirements, the certification strategy, and identifies the intended means of compliance.

> The Chief Airworthiness Engineer defines certification authority (EASA, FAA, cross certification, etc) and applies for Engine Type Certification.

> The Chief Engineer defines the means of compliance and identifies certification tests to be formally declared to, and witnessed by, the certifying authority through certification declaration and deviation reports.

Hand over ceremony for the Trent 500 following certification

Execution and reporting of tests phase:

> The Chief Engineer conducts all testing and analysis identified in the compliance.

> Development Engineering submits all documentation of compliance with requirements.

> The Chief Engineer confirms that the engine meets requirements through the statement of compliance to the authority.

In-service phase:

> The Chief Engineer undertakes the responsibilities of continued airworthiness of the product through its life-cycle.

In practice, only one aviation authority's requirements are adhered to during the certification process; cross certification is obtained through agreements between the main aviation authorities.

Approval of modifications to civil engine type design

All change packages, which will result in a modification, are subjected to assessment to

Engine							
70001		HP/IP Strain gauge		HP/IP Strain gauge			HP/IP
70002	Thrust reverser unit test			Cross wind and water ingestion			
70003			Operability	Bird strike			
70004	Sea-level pass off	Altitude performance and icing test		Engine pod leakage test	Functional test		
70005		Fleet leader IMI					
70006	150-hr		150-hr		Thrust reverser unit test		
70007		Fan blade-off test			150-hr		
71000	Sea-level pass off test for flying test bed	Flying test bed					

Certification

Engine manufacturer

Engine project	Airworthiness office

UK
Civil Aviation Authority (CAA)

US
Federal Aviation Authority (FAA)

EU
European Aviation Safety Authority (EASA)
Representing 36 countries including CAA (originally JAA)

The civil aerospace certification process

An example engine development plan showing how multiple engines are used to perform concurrent tests

establish verification needs. The subsequent certification or standardisation paperwork is completed in order to allow release of that modification for production and in-service engines by the Chief Engineer and aviation authority signing the modification bulletin.

Civil aviation certification testing
The 150-hour endurance test

The aviation authority requires that the integrity of the engine be demonstrated by the completion of the 150-hour endurance test. This test may be used for a number of purposes, including the demonstration of the integrity of a new engine or component design, or new operating limits.

The endurance test is a relatively short duration test of 150 hours in forward thrust but operating at conditions well beyond what will be encountered in service to give confidence to the aviation authorities that an engine of the design tested meets a certain mechanical standard; has satisfactory handling, functioning, and minimum performance; and is fit to enter service.

The 150-hour endurance test is desirable prior to flight trials and consists of a series of equal cycles containing running at maximum take-off and maximum continuous ratings, incremental/decremental running, and handling/running with and without off-take of bleed air.

On completion of the test, the engine is stripped, and the aviation authority is usually invited to view the hardware. Wherever possible, a formal layout with inspection details and data (for example, disc growths) is be provided. The endurance test establishes maximum values of parameters, such as shaft speeds and temperatures, for that particular build standard of engine; these must not be exceeded in service. The test does not purport to be a replica of the treatment the engine will get in service.

Fan-blade-off test

The aviation authority requires that the engine casings must be capable of containing the release of a single compressor or turbine blade, or any likely combinations of blades. The fan-blade-off test demonstrates mechanical integrity of all systems following the loss of a fan blade. It is a single-shot test,

comprising the explosive release of a fan blade where containment must be successful with minimal fluid system leakage. Certification may be confirmed with an engine test, a rig test, or analysis. The normal means of compliance for the fan blade is to demonstrate the containment of a fan blade, by deliberately releasing the portion of the blade outboard of the retention feature at the maximum LP shaft speed either on a full engine, or a fan-blade-off rig. The effect of the impact and subsequent run down on the gearbox and external units must be substantiated using the results of the test. The loads imparted to engine structure are analysed and reported.

Loads imparted to the airframe due to the event and the subsequent windmilling of the unbalanced fan must be agreed with the airframer as a specification issue. The release of core compressor and turbine blades is assessed by analysis of the potential radial release paths of each blade, and the containment capability of the casings in the release path.

Bird strike (foreign object) test

There is a series of tests to demonstrate the mechanical integrity of the engine following a bird strike event. Birds (dead, unfrozen, and of various weights) are fired at a running engine that must demonstrate acceptable operation following the strike, despite the resultant damage to the fan and core. The engine must not catch fire, burst, release dangerous fragments, or generate loads beyond the engine mount capabilities. The engine cannot lose the capability of being shut down, or create conditions hazardous to the aircraft.

The numbers, weight, and size of the birds are dependent on intake diameter. The fan

system is designed to cope with impact from a range of bird sizes at various positions on the fan face: the larger the diameter of the fan intake, the larger the weight of bird that must be accepted.

The following are typical certification tests:

› Large flocking bird ingestion – a 2kg (5lb) bird fired at a prescribed velocity when the engine is running at MTO (maximum take-off) thrust, and aimed at the most critical location on the engine face. The engine must maintain 50 per cent of MTO thrust and have the capability to continue at this thrust for 20 minutes after the bird damage.

› Large bird ingestion – an 3.6kg (8lb) bird is fired at a prescribed velocity at MTO thrust, aimed at the most critical location on the engine face, without power lever movement for 15 seconds after the event. The engine must be capable of shutting down safely and remain intact.

› Medium bird ingestion – four birds of 1.1kg (2lb) weight fired simultaneously at a critical velocity, at the most critical strike radii. The event is followed by a run-on period of approximately 20 minutes. The engine must not create hazardous aircraft conditions and still be able to produce 75 per cent of MTO thrust. A full engine test is required, with results extrapolated to worst day conditions.

› Small bird ingestion – one 0.25kg (0.5lb) bird is fired at the engine. Although this bird size arguably causes less mechanical damage than medium-sized birds, the debris could lodge undetected upstream of the fan, creating flow disturbance.

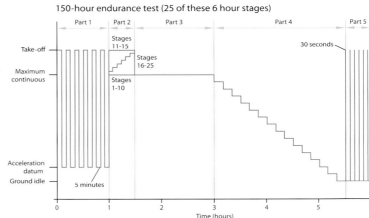

Typical cycles for completion of 150-hour test – note the variation in profile from stages 1-10, 11-15, and 16-25

44

Preparation for a fan-blade-off test – the brightly coloured blade is released with an explosive charge

Rain and hail ingestion test

The rain and hail ingestion testing simulates bad weather conditions and demonstrates that the engine can continue to operate in such situations. The requirement can be divided into three series of tests (identified as the most severe for engine operation in inclement weather):

〉 Rain low-power operability – the engine must operate acceptably during ingestion of certification standard concentrations of rain for three minutes. The engine must complete a cycle from flight idle to last bleed valve closure (nominated as the critical point) back to flight idle.

〉 Hail low-power operability – the engine must operate acceptably during ingestion of certification standard concentrations of hail for thirty seconds. The engine must complete a similar cycle as the rain low power operability.

〉 Rain high-power casing contraction – the engine must demonstrate that it does not suffer any unacceptable mechanical damage during and after the ingestion of rain at high engine power operations. The cycle consists of a stabilisation at MTO for three minutes to allow heat soaking, followed by the introduction of full certification standard concentrations of water within ten seconds, then three minutes of stabilisation with water on, followed by a rapid reduction in water within ten seconds.

For all of the rain and hail certification tests, the engine must typically demonstrate less than ten per cent performance loss during water ingestion and less than three per cent performance loss after water ingestion.

Altitude testing

Altitude testing is carried out to demonstrate the operability of the engine at altitude conditions. Installing the development test vehicle within an altitude test facility (ATF) simulates representative ambient temperature, pressure, and mass flows across the flight envelope. An altitude test facility can subject an engine to a wide variety of inlet temperatures and inlet and exhaust pressures, simulating the conditions it will encounter during aircraft operation.

Performance and functional tests demonstrate engine thrust, fuel consumption, acceleration and deceleration times, bleed air and power off-take capability, compressor surge margin, relight envelope, windmilling capability, capability to run with different fuels, reaction to control system failure, and oil system behaviour. The test is essential prior to installing the development engine onto a flying test bed (〉〉 47) or airframe manufacturer flight test vehicle for flight trials and certification.

Icing test

Icing testing demonstrates the mechanical integrity and operability of the engine during icing conditions (low temperature and high humidity). The engine is required to demonstrate its capability to function in those atmospheric conditions in which ice can form. The main threat is of ice building up on the static components at the front of the engine at low power, and then shedding into the engine en masse when the engine is accelerated. This can have a significant effect upon the temperatures, stability, tip clearance, and operation of the engine due to the sudden influx of cold matter and can result in mechanical damage from the mass of solid ice ingested.

The engine icing test is normally carried out at an ATF; a series of tests are run at a number of prescribed different altitudes and atmospheric liquid water content.

Rain ingestion testing with a full spray grid at sea level

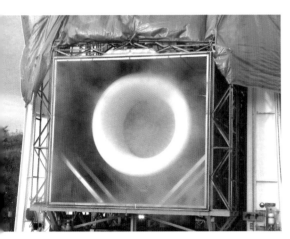

Low temperature starting test

During the test, ice is allowed to form and is shed by accelerating the engine to take-off power. The engine must not, as a result of the tests, have unacceptable increases in operating temperatures, immediate or ultimate reduction of engine performance, deterioration of handling characteristics, or mechanical damage.

Strain gauge and rotor integrity testing

Compressor and turbine rotor integrity can be established through rig testing replicating the following conditions:

> 125 per cent of red line speed (25 per cent above maximum normal engine speed).

> 110 per cent of the highest speed that could be reached due to failure.

> 105 per cent of the highest speed that could be reached due to failure of the most critical component or system or any other undetectable failure.

> Strain gauge testing is carried out with an engine test covering the range from idle to just above MTO. Strain gauges are placed on the blades and discs. The test is made as arduous as possible through the use of minimum component dimensions and lowest material properties within the likely manufacturing variation.

Low temperature starting test

The low temperature starting test demonstrates the mechanical integrity and start operability of the engine during low temperature conditions. The engine supplier must declare the minimum oil temperature for starting and also for accelerating from idle, while starting is completed with the minimum and maximum starting torque. Test evidence must show that the engine will accelerate smoothly, without engine damage, with the oil at the declared minimum temperature, when the throttle is moved from idle to MTO in one second or less. Ten starts are attempted and at least five must reach ground idle.

A declared cold starting test involves placing the engine in a cooling environment until the engine oil is at the temperature to be approved (-40°C is the normal target) and then attempting to start.

Cross-wind test

The cross-wind testing demonstrates the vibration characteristics of the fan and compressors when there is a wind blowing across the engine – proving that no unacceptable vibration resonance or fan flutter is exhibited.

With instrumentation on the compressor blades and/or the compressor discs, cross-winds of up to 83kmh (45 knots) are applied starting from head-on to the engine moving round to around 135° from engine centre line. Though not a certification requirement as part of the product development process, the engine is run at the highest possible speed with various cross-wind strengths.

Cyclic test

The engine used in this test is as close to the production standard as possible and the engine is run to a cycle devised to represent operating conditions. There are two typical cyclic tests carried out for certification:

> IMI (Initial Maintenance Inspection) – the number of cycles necessary to reach the first maintenance period.

> ETOPS and LROPS cycles – the number of cycles necessary to prove Extended Twin OPerationS or Long-Range OPerationS capability. (》 254)

The engine supplier also carries out a fleet leader programme, which consists of cyclic testing with enough cycles to stay ahead of the fleet leader – the operator who has flown the most cycles at any time. This test further demonstrates the mechanical integrity of the engine by catching any technical problems before an in-service incident.

Noise test

To determine the near and far field noise footprint of the engine, noise testing is completed using an open-air test facility. This is normally done in support of aircraft certification because noise is a whole aircraft issue. Microphones are situated in prescribed positions surrounding the engine to pick up noise signatures from intake, exhaust system, and bleed valve ducting.

Cross-wind testing

Boing 747 flying test bed used for Trent 800 engine testing

The nacelle standard must be representative of the in-service hardware (thrust reverser, nose cowl, fixed fan ducts, fan cowl doors, exhaust system nozzle) as this hardware directly impacts the noise signature generated.

The main non-engine hardware consists of a 'golf ball' or turbulence control screen. This special air-intake is designed to reduce intake distortion to a give cleaner intake flow to the engine. Another advantage of the golf ball is that the engine can operate on-condition at previously identified fan flutter avoidance zones with the elimination of cross-winds.

Use of the turbulence control screen on an outdoor rig with microphones arrayed around the engine

Flight test on new aircraft

Flight testing is used to demonstrate that the aircraft and engine combination is flight worthy prior to Entry Into Service (EIS) and covers a wide variety of engine issues like handling, relighting, zero-g oil test, performance, icing, and reverse thrust. For certification, noise is considered an aircraft issue, not an engine issue.

There are two main flight tests within the aircraft certification process:

> Flying test bed (FTB) – for engine certification, a previously certified aircraft flies with one new engine type installed. This may be an uncertified engine mark. (>> 238)

> Flight testing – having gained certification of the engine, a flight certification programme is carried out by the airframe manufacturer to gain aircraft certification. The first production engines of the newly certified engine mark normally support the flight certification programme.

The engine supplier obtains permission from the relevant aviation authority to approve modifications for use in flight tests.

Rig testing

The tests described so far have been whole engine tests, but much of the validation is performed on rig tests. This is not simply a function of cost. Rig tests are an essential precursor to whole engine testing: they can occur earlier in the process before a complete engine is ready, and, often, they can more easily accommodate the measuring and recording equipment used in the test. Validation of the combustor is carried out on rig tests for all of the above reasons.

Modelling and analysis

Wherever possible, the experimental approach prefers to validate a design through modelling and computer analysis rather than testing hardware. Applications such as finite element analysis are used to predict the structural integrity of a component. However, physical testing underpins such theoretical work; the success of modelling and analysis depends on the information used to create the model – information that comes from in-service data and engine and rig tests.

Defence aerospace military qualification

The qualification process for military engines begins with the identification of a requirement to validate a new military engine, to introduce a modification to an existing engine, or to amend the operating conditions of an existing engine. The defence department of the military customer is responsible for the airworthiness of the engine, taking the role that the civil aviation authority has for civil engine certification. An engine specification is agreed with the military customer; this includes both airworthiness requirements and all other requirements necessary to make the engine fit for purpose in the aircraft or weapon system. The engine specification does not distinguish between airworthiness and other requirements. The process of verifying that the engine meets all specification requirements is called 'Qualification'.

Qualification techniques

A number of different methods can used to substantiate compliance with the engine model specification. In general, they can be grouped into the following categories:

› Inspection – for example, the specification may prohibit the use of certain materials in the design of the product. In this case, compliance can be achieved by inspection of drawings.

› Demonstration – maintainability requirements such as the ability to change components on the engine can be shown to have been achieved by a demonstration on an engine.

› Analysis – demonstration of compliance with the model specification through analysis is very widely used. For example, whole engine finite element modelling is employed to confirm that there is no damaging resonance within the system. Often inputs to analysis models will need to be verified by test activity.

› Similarity – if the design concept of a component is similar to one previously qualified for a different application, it may be possible to use previously generated qualification evidence. In this instance, a

statement would be submitted to the airworthiness authority justifying the validity of this approach.

› Testing – can be further divided into component test, sub system test, or engine test.

Military qualification testing

There are seven key bench tests normally required by military customers:

› Endurance test (or Durability Proof Test).

› Accelerated Mission Endurance Test (AMET) – the engine is run on a sea level test bed to an endurance cycle representative of the intended aircraft missions and with the control system adjusted to represent normal in-service operation. The endurance cycle is

Preparing the 3 Bearing Swivel Module of the Joint Strike Fighter for testing

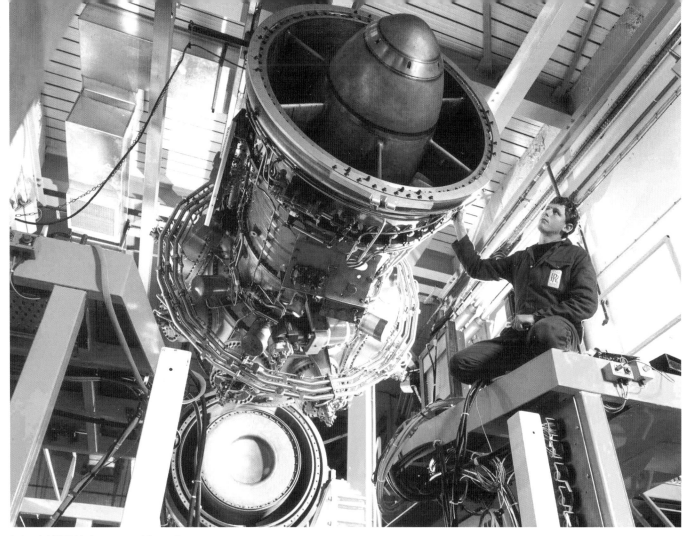

Industrial RB211 being prepared for testing

obtained by taking an 'average' mission and reducing the steady state running, leaving transients unchanged, so that the number of cycles between low power and high power normally occurring in one hour of service operation is compressed into a few minutes of testing. After completion of the test, the engine is fully stripped and inspected for damage. The test demonstrates engine life. The length of the test and, therefore, the life demonstrated depend on contractual agreements.

〉 Altitude test.

〉 Environmental icing test.

〉 Ingestion tests – similar to the civil aerospace bird strike, rain, and hail ingestion tests but with the possible addition of sand ingestion tests if analysis proves inadequate.

〉 Corrosion – the engine is run on a sea level test bed in a corrosive environment to determine the effect of corrosion in service. The test usually includes periods of running with salt solution continuously sprayed over the engine alternating with periods where the engine is stationary but at high humidity and temperature conditions to simulate storage.

〉 Exhaust smoke – the engine is run on a sea level test bed to allow the amount of smoke (unburnt carbon) from the exhaust to be measured.

There can also be tests to assess the engine's ability to withstand excess speed, temperature, and torque; these are similar to the strain gauge tests in civil aerospace.

Military flight testing

Flight testing may be required to ensure that the product is fit for purpose and to demonstrate compliance with the operational requirements of the relevant specifications for the engine and aircraft system. Flight conditions can be simulated in an altitude test facility (ATF), and while this testing is important for model validation and to provide evidence for allowing flight testing, flight testing itself offers the following further abilities:

〉 Testing in a fully representative installed environment that includes the effects of the intake, the aircraft forebody upstream of the intakes, the nacelle or engine bay (ventilation and cooling), and nozzle efflux entrainment.

〉 Testing in a fully representative aerodynamic environment that includes the effects of intake distortion due to normal aircraft operation and manoeuvring, formation flying involving rapid changes in flight conditions, natural icing, salt, and dust-laden conditions.

Compressor development rig

Industrial RB211 being installed

› Testing in a fully representative engine loading and systems integration environment that includes the effects of electrical, bleed air off-take, hydraulic loading during all aspects of engine and aircraft operation (steady state and transient), and cockpit displays.

› Testing for the effects of aircraft weapons firing and hot gas re-ingestion.

› Demonstrating installed engine operational characteristics to the customer or operator.

Energy gas turbine validation and verification

An aero-derivative industrial gas turbine is integrated into a package where it generally drives either an electrical generator, a gas compressor, or a pump for oil. A versatile product is required that can meet both the generic needs of these markets and specific customer requirements for operation, emissions, and fuel type. The design must meet the needs of onshore and offshore operators and differing regional legislation. These are the same regulations for any electrical equipment such as personal computers and mobile phones; the legislative bodies include CE (Europe), UL (USA) and CSA (Canada).

Energy gas turbine experimental process

The experimental requirements and process are similar in principle to those for aero engines. The gas turbine is not tested within a full package during the development programme, although before delivery the customer may request a 'string test' to prove systems integration. By the time this occurs, the experimental programme must have already ensured it will be a low risk, confirmatory test. To achieve this, in development the package is either modelled by the facility (for example, inertia to simulate loaded start) or represented by development hardware that meets the required physical and functional requirements (for example, fuel systems).

The Industrial RB211 powerplant removes the aero LP system and adds a free power turbine to harness the exhaust power. In the Industrial Trent powerplant, the fan is removed and a revised LP turbine powers both the driven equipment and an LP compressor upstream of the core. Each approach has some impact on the experimental programme, notably by allowing the RB211 to be tested as a gas generator exhausting through a nozzle rather than the free power turbine. This provides simplicity but adds an extra layer of systems integration; in recent times, a power turbine model has been included in the control system to allow a development engine to be controlled against simulated power turbine speed. When testing a Trent, the power generated by the turbine must always be absorbed, generally through load banks.

Within the core of the gas turbine, the most fundamental difference between an industrial and aero application is in the combustor. Industrial customers are usually able to choose from a range of combustors on the same gas turbine platform. All need to be proven experimentally.

Minimal changes are made to annular combustors taken from the aero programme, but industrial-specific injector designs are used for operation on liquid and gaseous fuels, including on-line fuel changeover. This expands the experimental programme as fuel system behaviour, hot end temperatures, starting, and operability data need to be acquired on both fuel types. If required, water injection to reduce emissions levels further adds to the experimental programme.

Validating dry low emissions (DLE) combustors adds an additional degree of complexity due to the ability to change the fuel splits between combustor zones. Functional testing is carried out to optimise emissions and noise. Compliance data is then acquired to establish emissions guarantee margins and the operating map.

Industrial gas turbine designs maintain a high level of component commonality with the aero engine. The data generated by the aerospace experimental programme is, therefore, applied to the industrial product to reduce the cost and time of research and development. The industrial development programme assesses what can be carried over from the aerospace development and what new validation is required.

Typical industrial specific testing includes

> endurance
> DLE combustion
> DLE controls.

Maximum operating temperatures in industrial applications are lower than in aerospace, but higher hot end temperatures during continuous running result in component lives being dominated by creep, oxidation, and sulphidation. The operating regime also contributes to this – gas compression units operate predominately at high power and accumulate cycles slowly. Only units in power generation, particularly peak lopping applications, accumulate cycles with regularity. For these reasons, cyclic endurance tests are not typically carried out although for any given product there is usually some form of endurance test run against a typical customer operating profile.

The principles and processes of an industrial gas turbine experimental programme do not differ significantly from aerospace, although it does take account of design differences at system, subsystem, and component level.

Marine and naval qualification
For marine gas turbine qualification, the new engine type must meet the following conditions, and the EDP is constructed to meet both requirements:

> Company design and validations standards – this process is very similar to the aero validation rules but ingestion requirements are replaced by shock requirements.

The Type 45 frigate for the Royal Navy

> External industry standards – for commercial applications, this means 'Classification Society Rules'. Historically for naval applications, the customer, for example the Royal Navy or US Navy, certified the application. Increasingly, however, they employ a classification society.

Marine classification society
A classification society is the marine equivalent of an airworthiness authority. The classification society approval of a marine product means that it meets the international legal standards for safety of life at sea and is recognised by marine vessel insurers as acceptable.

There are four major international class societies:

> Lloyds Register, based in the UK
> Det Norske Veritas (DnV), based in Norway
> Bureau Veritas, based in France
> American Bureau of Shipping (ABS), based in the USA

Marine validation strategy
As the marine gas turbine technology is taken from existing aero and industrial engines, it can take advantage of proven engineering databases for many components. Some class societies and the US Navy have accepted read-across from existing aero and industrial certification databases.

However, certain key tests have to be completed on the marine product that are outside the scope of both aero and industrial development programmes where fundamental differences occur between the marine product and the aero equivalent. Such differences include the power off-take shaft, control system, component design changes, cycle changes

affecting secondary air system, whole engine dynamics and bearing systems, oil system changes associated with the deletion of LP shaft and subsequent bearing load changes, and the marine operating regime.

Typical marine tests include

> functional testing
> start testing
> gas turbine alternator testing – loadshed and 0-50-100 per cent stepped load test
> piping integrity test
> vibration survey
> airborne noise
> rotor integrity (a significant overspeed test for 'developmental' engines – not applicable to aero-derivatives such as the MT30)
> overspeed and trip demonstration
> endurance test – typically 1,500 hours
> shock tests
> sea trials.

Typical classification society naval certification programme
The major elements to a naval certification programme are

> Design assessment
> Fabrication of First Article engine – with the inspection witnessed by the relevant class society
> First Article testing
> Sea trials – the class society will issue the Machinery or Type Certificate on completion of sea trials.

In the 21st century, there is one fundamental design challenge for the jet engine: how to maximise its benefit to mankind without damaging the fragile world around us.

environmental impact

A CENTURY AGO, MOTORISED TRANSPORT WAS PROMOTED AS BEING EMISSION-FREE – UNLIKE THE HORSE IT WAS REPLACING; IN THE 1920s, A FLEDGLING AND FRAGILE AERO INDUSTRY WAS SHIELDED FROM COMPLAINTS ABOUT NOISE. TODAY, A MORE EXPERIENCED WORLD RECOGNISES THAT REDUCING ENVIRONMENTAL IMPACT IS ONE OF THE MAJOR ENGINEERING CHALLENGES OF THE 21st CENTURY.

environmental impact

In recent years, two design requirements have received a high priority from customers and engine manufacturers alike: the reduction of noise and the reduction of emissions, arguably the two least wanted by-products of the gas turbine – as they are of many industrial processes and modern forms of transport.

Considerable research and development is going into the reduction of these by-products and significant improvements have been achieved. However, customer requirements are becoming ever more challenging and much remains to be done.

Noise

Modern aircraft are significantly quieter than earlier designs, with reducing specific thrust (or increasing bypass ratio) being an important contributory factor via lower jet velocities. Modern aircraft emit only one per cent of the sound energy emitted by aircraft designed forty years ago. However, continued environmental pressures for further reductions make noise control one of the most important fields of aero engine research.

Noise control

Aircraft are regulated using standards set by the International Civil Aviation Organization (ICAO). There are three reference locations at which the noise limits are specified: two for take-off (lateral and flyover); and one for landing (approach).

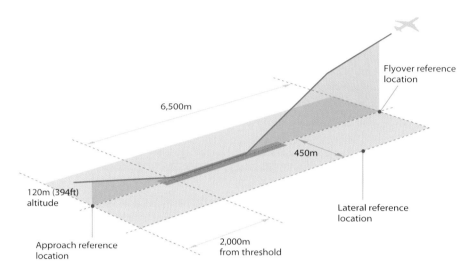

Noise certification reference locations as specified by ICAO

In each case, the noise is measured during take-off or landing and is expressed in terms of the Effective Perceived Noise Level (EPNL), a decibel unit that takes into account the frequency, content, and duration of the event.

The current statutory noise certification requirements were agreed in 1977 and are known as ICAO Annex 16 Chapter 3 or simply 'Chapter 3'. The member states of ICAO have adopted this in their individual national legislation, the most commonly known example being the US Federal Aviation Regulation (FAR) Part 36 Stage 3, which is virtually identical to Chapter 3. The maximum permitted noise is dependent upon the design weight of the aircraft. As a general rule, the noise limit increases as aircraft weight increases, but there is a plateau at low and high aircraft weights.

Effective from 1 January 2006, 'Chapter 4' sets more stringent requirements for the certification of new aircraft types. This requires a noise level cumulatively 10EPNdB (effective perceived noise in decibels) below the cumulative Chapter 3 limit, in addition to meeting other conditions.

A cumulative margin of 10EPNdB means that the sum of the lateral, flyover, and approach noise levels must be at least 10EPNdB below the summed Chapter 3 noise limit at those three conditions. In addition, the Chapter 3 limit cannot be exceeded at any condition

and there must be a cumulative margin of at least 2EPNdB against Chapter 3 for any two conditions.

In addition to the international requirements, some airports have even more stringent restrictions on noise levels. The number of airports that have their own individual noise requirements has rapidly expanded in recent years. In some cases, there are various operational restrictions on aircraft that exceed the airport-prescribed noise levels; in other cases, there are noise-related landing fees or even fines if the measured noise level is too

high. One of the best known examples of a local airport rule is that at the London Airports Heathrow, Gatwick, and Stanstead, where the combination of high air traffic volumes and high population density have led to limits on departure and landing noise. The Quota Count system was introduced to control night-time noise and, unlike the ICAO limits, does not give any alleviation for aircraft size. This has meant that the Quota Count requirements are much more demanding than the ICAO limits for large aircraft types and so have driven recent noise technology requirements.

Progress in noise reduction

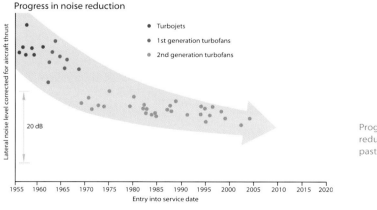

Progress in noise reduction over the past 50 years

Maximum permitted and achieved noise levels

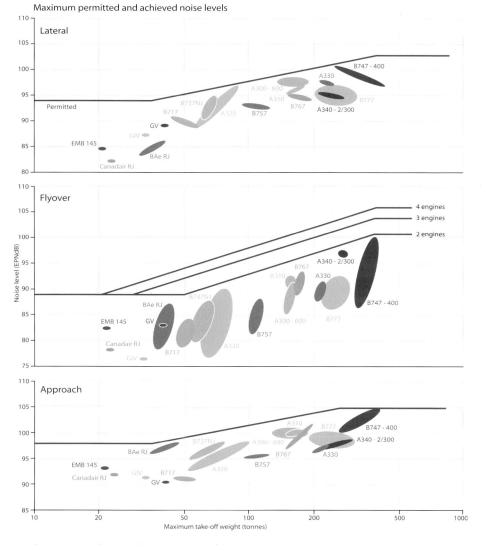

Statutory noise limits and some examples of the achieved noise performance for a range of aircraft types. There is a wide range of achieved noise levels, caused by the many different weights and engine thrust ratings for some aircraft types.

The main contributing noise sources for take-off and approach: fan, compressor, combustor, turbine, jet, and airframe

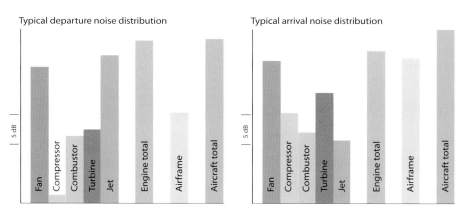

Sources of aircraft noise

The sound heard from an aircraft is, in fact, the result of many individual and quite separate noise sources added together. This is not a straightforward addition – for example, two sources with a noise level of 100dB each will add to an overall noise level of 103dB, because the ear perceives noise as a logarithmic function of power.

Both engine design and airframe characteristics greatly affect the operational noise levels of aircraft. For example, improved airframe aerodynamic performance can reduce the maximum thrust required and also allow the aircraft to climb away rapidly from the population (sound pressure decreases with distance from the source). For this and other reasons, aircraft noise control is a highly integrated activity between the aircraft and engine manufacturer.

The relative values of the main constituent noise sources can vary significantly from case to case, but some general observations can be made about how the relative importance of the noise sources varies between the three certification conditions. For example, jet noise is the most important source at the lateral condition, where full engine thrust is required, but it is well below the fan and airframe noise levels at the approach condition, because the engine is throttled back during the descent. As a result, the contribution from the aerodynamic disturbances, created by the aircraft undercarriage and lifting and control surfaces such as flaps and slats, becomes very important during approach. At the flyover condition, a reduced take-off thrust is selected at a safe altitude to abate noise, resulting in fan and jet noise sources both being important in setting the received noise on the ground. Other noise sources such as the low pressure (LP) turbine, combustor, and compressor can also add to the total signature.

Noise of a typical 1960s engine

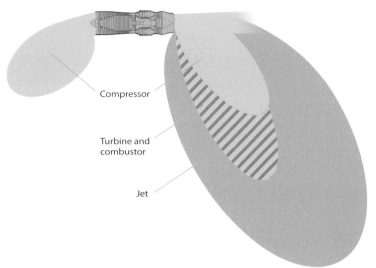

Compressor

Turbine and combustor

Jet

Noise of a typical 1990s engine

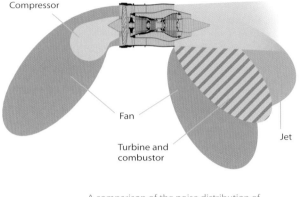

Compressor

Fan

Jet

Turbine and combustor

A comparison of the noise distribution of two generations of engines. The bubbles approximately indicate the relative 'size' of the main individual engine noise sources and the angular extent indicates where each is most prominent. The noise contributions from a modern turbofan engine are greatly reduced and much more evenly matched than from a turbojet.

Though the noise contributions are now much more evenly matched, fan noise has emerged as a very important source.

Further progress in reducing the aircraft noise level is only possible if all of the important constituent sources are reduced – this is because, as stated, with the decibel scale reflecting the response of the human ear, the constituent sources add logarithmically, not algebraically.

The decibel unit is used because sound consists of pressure fluctuations and the human ear can detect a very wide range of amplitudes. The human ear can usually distinguish between signals 3dB apart, but cannot reliably notice smaller changes. It is interesting to note that engine noise prediction and measurement techniques require accuracies significantly better than 3dB.

The problem is complex because there is usually not a single source of noise that can be 'fixed'; instead several sources, often with very different control measures, must be tackled in order to make significant progress in reducing the noise signal around airports.

Fan noise

The fan system produces a noise field that is perhaps the most complicated in the engine. Noise comes from the fan blade aerodynamics and the outlet guide vanes, as well as aerodynamic interaction between the blades and vanes. The numbers of fan blades and outlet guide vanes, and the gap between blades and vanes, affect how much noise is created. The noise produced by the fan system passes up the intake duct and then radiates out of the intake into the atmosphere. It also passes down the length of the bypass duct and radiates out of

the cold jet nozzle into the atmosphere. The fan system noise is made up of two, very different, types of sound: broadband and tone.

Broadband noise

Broadband noise sounds like a hiss. An example of broadband noise is the sound heard inside a car when travelling quickly on a motorway.

Broadband noise is made up of many different frequencies. The fan system broadband noise comes from the turbulent air in the boundary layer near the surface of the aerofoils and in the wakes behind the fan blades and outlet guide vanes. The noise is generated in exactly the same way as by the car on the motorway. The more aerodynamically efficient the fan blades are, the less broadband noise is generated: similarly, the more streamlined the car body shape, the quieter the car interior.

Comparison of car with a fan blade

The similarity between broadband noise from a jet engine's fan and that of a car on the motorway

Pressure fluctuations in the boundary layer and trailing air (wake) create a broadband noise

Tone noise

Tone noise sounds like a whistle, the hum of a refrigerator, or a noisy two-stroke motorbike. It is sound energy concentrated in just one frequency.

The pressure wave just in front of each fan blade produces a sound pulse each time a blade goes past. These pressure waves produce tone noise at the blade passing frequency – the revolution rate per second multiplied by the number of fan blades.

Tone noise gets much louder when the fan blade tips reach supersonic speeds. The pressure waves make the air in the engine intake resonate (like a string vibrating). If the speed of the pressure waves is high enough, a large amount of energy flows along the intake and out of the front of the engine. Acoustics engineers describe a sound as 'cut-on' when a large amount of energy is flowing. Careful shaping of the fan blades can reduce the amount of tone noise generated from the fan system – swept fan blades offer big tone noise reductions.

Another type of tone noise generated by the fan blades is called 'buzz' because it sounds like a buzz, or circular, saw cutting wood. This noise is made up of a collection of regularly spaced tones, and is often heard inside an aircraft fuselage during take-off.

A Fourier decomposition shows how much of each frequency is present in a sound. A decomposition for a saxophone and violin would show why they sound very different even when playing the same note. Each produces a very complicated, and very different, collection of frequencies. A tuning fork is one of the few mechanical devices that produce virtually a pure tone of just one frequency. In a Fourier decomposition, a tuning fork would look like a single spike because all the noise is at one frequency; on the other hand, car interior noise on the motorway would look like a flat horizontal line because it contains many frequencies.

Frequency decompositions of fan noise at approach and at take-off. As well as being louder at take-off, the 'shape' of the noise can be seen to be very different at the two conditions.

Frequency decomposition of fan noise at approach

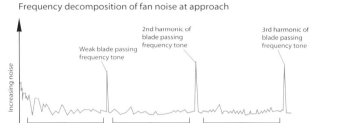

Frequency decomposition of fan noise at take-off

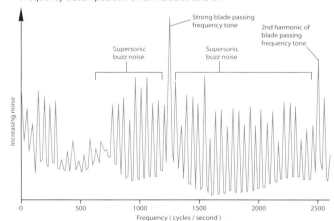

Slice through CFD solution showing pressure wave fronts travelling forwards from fan blades

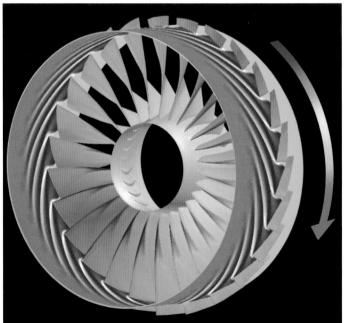

A CFD prediction of buzz noise being generated by the fan blades

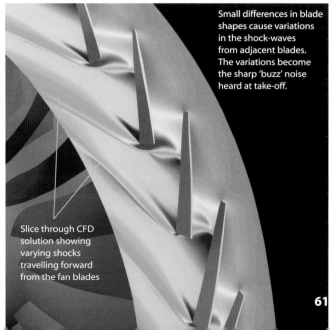

Small differences in blade shapes cause variations in the shock-waves from adjacent blades. The variations become the sharp 'buzz' noise heard at take-off.

Slice through CFD solution showing varying shocks travelling forward from the fan blades

61

A model fan rig in a test facility – the anechoic chamber at the Anecom facility in Germany. The facility floor area is 37m by 31m and the ceiling is 10m high. The model fan shown has a diameter of just under one metre.

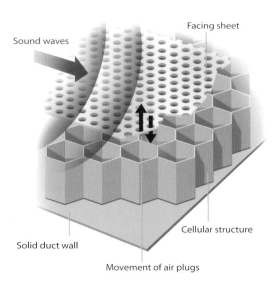

Structure and working principles of acoustic panels. Energy is absorbed through frictional dissipation into heat.

Sound waves

Facing sheet

Cellular structure

Solid duct wall

Movement of air plugs

The buzz noise is only created when the fan blades are rotating at supersonic speeds. When the blades are moving supersonically, there are aerodynamic shocks in the passages between the blades. These shocks are very similar to the sonic booms produced by supersonic aircraft. Very slight differences in the manufactured shapes of the different fan blades in the fan set give rise to differences in the passage shock shapes. It is these shape variations that cause the buzz noise to be produced.

Careful design of the fan geometry can reduce buzz noise. Also, designing the fan to rotate more slowly at take-off reduces the shock strength and subsequent buzz noise.

Fan noise testing

The noise the fan system makes (on its own) is measured by running a model of the fan system in a special quiet chamber known as an anechoic chamber.

For practicality, these rigs are smaller than the actual engine components; it is possible to accurately scale the results to full size because the scaling effects are well understood (for example, tone frequencies from rotors simply scale with rpm). The amount of instrumentation can be very extensive, sometimes involving several hundred microphones inside and around the rig to measure the noise, examining how sound is generated and how it propagates out of the engine.

Reducing fan noise

Another important way of reducing noise levels is to absorb the sound energy after it has been produced. On modern jet engines, the intake and bypass duct are lined with special panels that absorb the sound produced by the fan system. Similar panels can be found at the side of roads that pass through built-up areas. These acoustic panels work by resonating to the sound energy, and then dissipating the energy as heat into the air. On industrial and marine installations, the enclosures are acoustically treated in this way.

Controlling how much fan noise escapes forward out of the engine can also be achieved by careful shaping of the engine intake geometry. A 'scarfed' intake is shaped to deflect the acoustic energy upwards, away from any community below the aircraft.

Exhaust jet noise

The exhaust jet is the principal source of noise when the engine is operating at full power during aircraft take-off. At the high thrust setting required under these conditions, the exhaust gases are expelled from the nozzle at high velocity, and noise is generated by the turbulent mixing of these gases with the surrounding air. The magnitude of the turbulence is proportional to the velocity difference between the exhaust gases and their surroundings. This velocity difference is known as the velocity shear. The principal controlling parameter is, therefore, the mean velocity of the jet. The noise of a single-stream jet increases with the eighth power of the velocity, a result predicted by theoretical modelling in the 1950s and validated by test experience. It is known as the V^8 law.

Noise reflected away from ground by extended lower lip

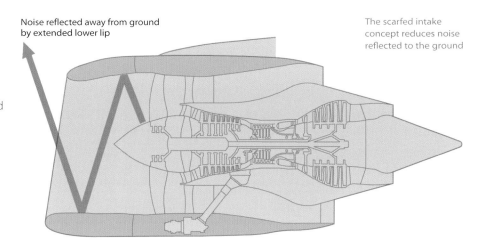

The scarfed intake concept reduces noise reflected to the ground

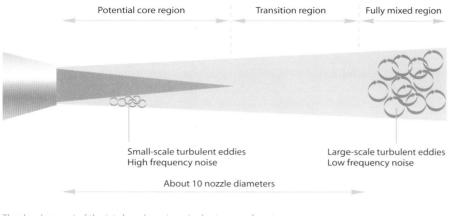

Potential core region | Transition region | Fully mixed region

Small-scale turbulent eddies
High frequency noise

Large-scale turbulent eddies
Low frequency noise

About 10 nozzle diameters

The development of the jet shear layer in a single stream exhaust

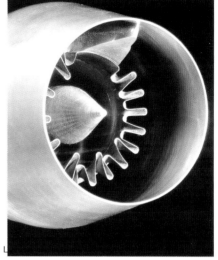

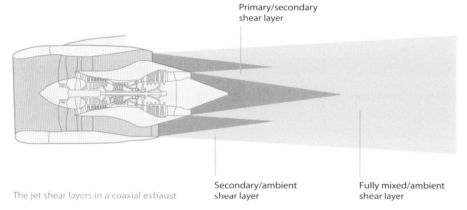

Primary/secondary
shear layer

The jet shear layers in a coaxial exhaust

Secondary/ambient
shear layer

Fully mixed/ambient
shear layer

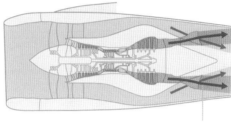

Lobes direct core and
bypass flows together

Model of a lobed core mixer

Jet noise is unique among engine noise sources in that it is generated outside the engine. The mixing process and the noise generation takes place over a considerable axial distance, up to ten nozzle diameters or more downstream of the engine. As the jet develops in the downstream direction, the lengthscale, or size, of the turbulence in the annular mixing layer increases.

Higher frequency noise is generated close to the nozzle exit due to the smaller lengthscale of the fluctuations; lower frequency noise is generated further downstream where the lengthscale of the turbulent fluctuations ultimately becomes comparable with the jet diameter. The general principle also applies to two-stream or coaxial jets, but the situation is more complex because of the additional shear layers.

Historically, jet mixing noise reductions have gone hand-in-hand with reductions in specific thrust and increases in bypass ratio as a result of the lower mean jet velocity

required to achieve a given thrust level. The addition of a slower-moving, secondary, or bypass, stream of air exhausting concentrically around the primary, or core, jet results in two annular mixing regions with significantly lower shear than that created by a single-stream jet at the same thrust.

At moderate bypass ratios of less than five to one, further jet noise reductions can be realised by mixing the core and bypass streams before exhausting the total flow to atmosphere. The mixing process can be enhanced by using a lobed core mixer, but for appreciable noise reduction the required duct length can still be quite large (around two nozzle diameters). Consequently, the drag and weight penalties of a long cowl bypass nozzle, plus the convoluted mixer, need to be considered to determine if this is the optimum nozzle configuration for a particular aircraft application.

In recent years, jet noise reductions have been sought by means of nozzle serrations. The enhanced mixing produced by the

serrations can result in small but significant jet noise benefits (with acceptable aerodynamic performance); several production engine applications have been identified.

The velocity shear effect, referred to above in the context of coaxial jets, also features in the jet noise change between static and flight operation of the engine. When the aircraft has forward speed, the velocity shear between the exhaust gases and the atmosphere is reduced, and the jet noise can reduce by, typically, five to ten decibels. In order to understand these very large 'flight effects', experimental testing is often carried out in purpose-designed anechoic chambers.

A 20-25cm diameter model of an engine nozzle, tested at the actual jet velocities and temperatures experienced by the engine, can be scaled in frequency and intensity to give very close agreement with the full size engine (eight to ten times larger). Using such facilities allows various designs to be evaluated, and noise reductions (determined with

A model of a serrated nacelle and core in an anechoic chamber

simulated aircraft forward speed) applied without costly full-scale testing.

Due to the distributed nature of the jet flow and its associated noise, acoustic and aerodynamic interactions with the airframe structure need to be considered. For example, in common with other rear arc sources, jet noise can be reflected off the wing, but the proximity of the flow to the wing and even the flow scrubbing the wing surface (when the wing flaps are deployed) can cause noise. Future aircraft applications might achieve reductions if the engine and aircraft can be integrated in a way that reduces or eliminates these effects.

LP turbine noise

The HP and IP turbines tend not to be important sources of noise because they are buried in the core and so their noise is contained within the engine. The LP turbine, however, does require noise control, which is often achieved using similar principles to the fan system. As with the fan, tone noise can be trapped within the engine by selecting the aerofoil numbers to achieve acoustic 'cut-off'. It is possible to exploit the fact that the human ear is less sensitive to frequencies above about 4kHz by choosing the rotor

numbers to generate sound only at these frequencies (which are also attenuated more by the atmosphere). The multi-stage design of the turbine means that the most appropriate combination of noise control features is often the result of iterative noise and aerodynamic studies to get the optimum configuration.

Combustor noise

On most engine designs, the noise contribution from the combustion process is not significant at the noise certification flight conditions. Noise created by instabilities in the combustor, for example during start-up, is controlled by air/fuel ratio management. This becomes more difficult for lean, low emission combustor designs. The ultra low emission designs necessary in some land-based industrial applications require additional forms of 'noise' control. (» 126, 127)

Secondary systems

As progress is made in reducing the primary sources of aircraft and engine noise, the contribution from secondary features such as off-takes or exhaust ports can become important. Noise control is now often part of the design requirement for secondary systems.

Aircraft and engine noise testing

The accurate measurement of aircraft and engine noise requires a carefully controlled experimental set-up. For example, the measured sound is greatly affected by the atmospheric conditions, and so the ICAO certification requirements stipulate strict wind limits plus correction factors for temperature and relative humidity to account for the atmospheric attenuation of sound. Another example is that the requirement to reproduce the in-flight inlet fan noise leads to the need to remove atmospheric turbulence during ground testing; this is achieved by using a large, yet acoustically transparent, air filtering device known as a turbulence control screen.

The structure consists of individual flat panels, with a perforated face sheet and supporting honeycomb, giving it an appearance that leads to the common name of a noise 'golf ball'.

A procedure has been developed known as the Noise Family Plan that allows noise certification of derivative engines to be achieved by building upon a read-across between ground and flight noise testing. By developing this read-across for the so-called 'parent' aircraft and engine combination, the noise certification of subsequent engine derivatives in that family can be achieved by ground testing alone. Indeed, this process is so well established that ground noise tests are regularly used during research and development programmes to give a very good indication of the eventual in-flight noise levels.

Part of this validation testing work involves deployment of many microphones to allow detailed diagnostic investigation of noise generation at source, the effect of the acoustic treatment, and propagation along (and radiation from) the nacelle ducting. This may involve hundreds of microphones inside and outside the engine. Increasingly, advanced array designs are used in conjunction with phase-related signal processing to determine features such the modal composition and spatial distribution of individual noise sources.

In addition to ground testing, flight tests are important to aid the development of some noise technologies. These programmes are often large and expensive but are also often ideal opportunities for collaboration because many noise solutions involve the integration of ideas from the engine, aircraft, and nacelle.

Continued research

For several decades, there has been sustained research, enabling dramatic reductions in aircraft noise. More recently large collaborative programmes have been launched, bringing together aircraft and engine manufacturers and key members of the supply chain to provide a holistic approach to noise reduction. The combined expenditure of these programmes runs into hundreds of millions of pounds.

Airlines, airports, manufacturers, and air navigation service providers need to apply a balanced approach to noise management around airports. This comprises reduction of noise at source, land-use planning, noise abatement procedures, and operating restrictions, with the goal of addressing the local noise challenge in the most cost-effective manner.

Manufacturers need to develop and promote new technology to reduce aircraft engine noise consistent with emissions and fuel efficiency needs. Research goals are aimed at reducing perceived take-off and landing noise by 50 per cent (10dB) by 2020 from levels in 2000. This will involve novel engine and aircraft architectures in addition to developments in low-noise technology.

The acoustically transparent turbulence control screen used to remove atmospheric turbulence

Some of the microphone instrumentation around an engine during a ground noise test

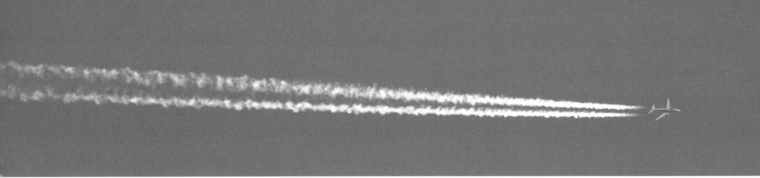

Emissions

Emissions from a gas turbine are a result of the combustion process, and it is in the combustor that major developments are being introduced in order to counter the environmental impact of those emissions. Much work is being done on controlling the air/fuel ratios and temperatures of the gases at different points of the combustion cycle. (» 126) Industrial applications, which have had to face more stringent regulation and which do not have the weight and space constraints of aero engines, are demonstrating new approaches to emissions control. (» 127)

Gas turbines, emissions, and the environment

There is increasing concern about the accumulation of man-made greenhouse gases in the atmosphere leading to increased risk of climate change. The predominant man-made greenhouse gas is carbon dioxide, which is released into the atmosphere when fossil fuels are burnt. CO_2 emissions from gas turbines can be reduced by using fuels with a lower carbon content, and by increasing the efficiency of the engine.

The efficiency of the gas turbine compares favourably with other types of power and the gas turbine's ability to run on natural gas, which has a low carbon content compared to coal, has made it attractive for land-based power generation. Efficiency can be further improved by using some of the exhaust heat to drive a steam turbine in combined cycle mode, or, where there is a demand for the heat energy, in combined heat and power plants.

Aircraft emissions are of particular concern to the global environment due to the altitude at which they are omitted. Water is an inevitable combustion product. Oxides of nitrogen, NO_x, are generated due to the very high temperatures and pressures in the combustor, leading to dissociation and reaction of the nitrogen and oxygen in the air. Leaner combustion processes

reduce temperatures and, therefore, NO_x formation; they also reduce the generation of soot particles, which may contribute to contrail formation. Industrial applications lead in the implementation of such technologies due to much lower emissions requirements for static plant in areas of human habitation – but focus is also maintained on local air quality in the vicitinity of airports.

Specifically, for aero engines, weight is a significant environmental performance issue, as reduced weight will contribute to the overall aircraft performance, leading to lower thrust requirements and therefore reduced fuel burn, emissions, and noise.

In addition to climate change, marine engines have to address issues associated with the sensitive marine environment and air quality around ports while operating on marine diesel fuels. Increased use of gas turbines in place of traditional marine engines could help the marine industry to tackle these problems.

Increasingly there are trade-offs in the design of gas turbine engines between global issues such as fuel use and climate change and local issues such as noxious emissions and noise.

Gas turbines are manufactured using a range of materials, some of which are specialised, rare, and highly processed. Also, the manufacture can involve processes and substances that are hazardous to humans and the environment. While every effort is made to reduce or avoid these circumstances, an understanding of the environmental impact of the whole life-cycle of the engine can show that, in many cases, carefully controlled use of some hazardous materials and processes can be acceptable because of their beneficial effect on the overall performance and impact of the engine.

The environmental life-cycle of a gas turbine

All products can be said to follow a 'life-cycle'. At each step, material is used that will form

part of the product: consumables such as coolants and cleaning fluids; resources such as electricity, gas, and oil. And, at each step, waste arises from scrapped parts, packaging, waste water, chemicals, and air emissions. All of these, arising from the product's life-cycle, cause environmental impacts.

Environmental life-cycle studies of gas turbines have shown that the biggest environmental impacts are caused by consumption of fuel and the emission of gases during the use of the turbine. The major impacts are as follows:

> global warming from CO_2, H_2O, and contrails

> acid rain and health risks from NO_x, CO, and unburnt hydrocarbons (UHCs)

> acid rain and global warming from SO_x

> health risks and global warming from particulate matter.

Customers are in turn affected by these impacts with operational restrictions, direct fuel costs, and with problems obtaining planning permission for airports and power plants.

Throughout the engine's life-cycle, customers and gas turbine manufacturers also manage increasing costs of raw materials, energy, and waste disposal. The most effective way of managing costs, risks, and environmental impact of products is to make environmental considerations a fundamental part of the decision making during the design process.

Consequently, almost all new designs must reduce the environmental impact of the gas turbine, with particular emphasis on fuel use, but also considering all other life-cycle stages.

Climate change

Gas turbines traditionally consume fossil fuels and emit the combustion products directly to the atmosphere. This contributes to the accumulation of greenhouse gases in the atmosphere, believed by the majority of world climate experts to be contributing to man-made climate change.

CO_2 and other emissions from gas turbines such as water vapour, oxides of nitrogen, unburnt hydrocarbons, and particulate matter have varying effects depending on the location of the emissions. At ground level, these emissions have only local or regional effects, but aero engine emissions at altitude can have a significant impact on the global atmosphere, making an additional contribution to climate change. The scientific understanding of this phenomenon, which includes the creation of ozone, destruction of methane, and the impact of contrails and cirrus clouds is currently poor, and is receiving much attention from the research community.

Emission species
Carbon dioxide (CO_2)
This is believed to be the main atmospheric gas contributing to global warming. It is a product of complete combustion of hydrocarbon fuel. Therefore, as it is directly related to fuel burn rate and is an unavoidable by-product of combustion, it cannot be reduced directly by

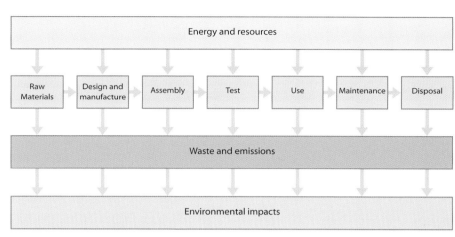

Environmental product life-cycle

combustor design. Control of carbon dioxide emissions has to be achieved through improving overall engine and airframe efficiencies.

Water Vapour (H_2O)
Water vapour, like CO_2, is a product of complete combustion and is not important in the troposphere where the air is still humid; under these circumstances, there is limited global warming potential. However, water

vapour from supersonic aircraft is a strong global warming agent in the stratosphere, where the air is too dry for contrails. When water vapour is visible in the exhaust it is usually referred to as a contrail.

Contrails
All aero engines emit an invisible stream of aerosols and condensable gases, such as H_2O (water vapour), and H_2SO_4 (sulphuric acid), which lead to the formation of new volatile liquid particles. The formation of these particles depends on the mixing of the exhaust gases with the ambient air, the plume cooling rate, and the plume chemistry. In addition to the volatile liquids, non-volatile solids such as soot particles formed during combustion are present in exhaust plumes.

Under certain thermodynamic conditions, the water vapour freezes to form ice particles causing the formation of a condensation trail, or contrail. The main controlling factor is the relative humidity in the plume that results from the mixing of the warm moist gases of the exhaust with the colder, less humid, surrounding air. For contrails to form, the relative humidity of the young plume must be 100 per cent. Contrail ice particles nucleate mainly on the soot and volatile sulphur particles found in the exhaust plume.

Contrails will rapidly disappear after the passage of the aircraft if the ambient humidity is low. However, if the humidity of the atmosphere is above ice saturation, these clouds can persist and grow through continued deposition of ambient water.

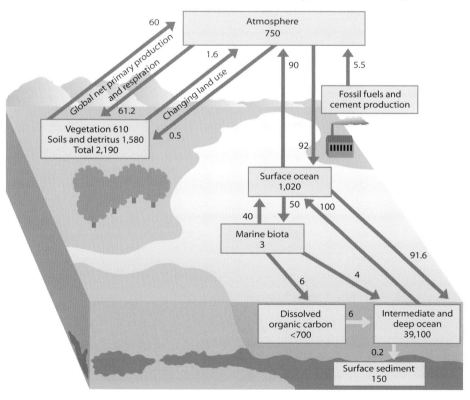

The carbon cycle. Carbon is in a perpetual loop, changing form and function depending on its location in the cycle.

Typical contrail formation over northern Europe

Typical contrail formation over northern Europe

It is believed that contrails would still form even if all the particles in the exhaust were removed, as the water would continue to condense on existing atmospheric particles. Recent studies have shown an increase in cirrus clouds in areas of high aircraft activity.

Oxides of nitrogen (NO$_x$)

NO$_x$ is mainly made up of NO and NO$_2$. It is predominately produced through the consumption of fossil fuels, and so the major sources are heavy industry, transport, and power stations. Figures obtained through a report carried by environmental agencies during 1992 suggests that 82 per cent of NO$_x$ emissions are created by road traffic and power stations, only two to three per cent by aircraft. There are three types of NO$_x$ formed during the combustion process:

> fuel NO$_x$ – comes from nitrogen being oxidised by combustion air

> thermal NO$_x$ – generated by nitrogen reacting with a surplus of oxygen at high temperatures

> prompt NO$_x$ – results from the formation of hydrogen cyanide (HCN) then oxidising to form nitric oxide (NO).

NO$_x$ can be carried for long distances causing health risks and contributing to acid rain. It is a source of ozone production in the troposphere adding to global warming, while depleting ozone in the upper stratosphere where this atmospheric gas filters out some of the sun's harmful rays. NO$_x$ can also form photochemical smog at ground level. Visible NO$_x$ is an important issue, especially for marine applications. NO$_2$ is a visible brown gas; a major concern is its concentration leaving the exhaust stack – and whether it is below the threshold of visibility.

Carbon monoxide (CO)

Carbon monoxide is a poisonous gas and is a product of incomplete combustion. This is a low-power issue for aircraft engines. The main producer of this gas is road transport (around 90 per cent).

Unburnt hydrocarbons (UHC)

Unburnt hydrocarbons contribute to photochemical smog, in addition to acid rain and health problems. The majority of UHC production is by road traffic and solvent evaporation. In a gas turbine, UHC is produced as a product of incomplete combustion due to low pressure and low gas temperatures in the combustor; it is generally, therefore, a low-power problem like carbon monoxide. Its presence reduces as power is increased above idle and no UHC is produced at most flight conditions.

Oxides of sulphur (SO$_x$)

Oxides of sulphur add to the problem of acid rain, but limits imposed on the quantity of sulphur in aviation fuel control the output of SO$_x$ from the aero engine. The average fuel only contains from 0.04 to 0.05 per cent of

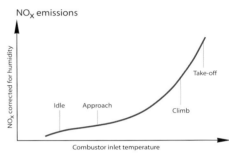

Typical NO$_x$ emission characteristic of a jet engine combustor. The highest emissions are at take-off, where the high temperatures within the combustor cause the nitrogen and oxygen in the air to combine.

This is an aviation NO$_x$ inventory for the 1992 fleet. NO$_x$ reactions are dependent on altitude and sunlight so global NO$_x$ inventories for aviation are completed in three dimensions. The emissions are averaged over the complete longitude at each latitude. The North Atlantic flight corridor can clearly be seen at a latitude of +30°, highlighted by the dark red patches (signalling a large intensity). This region would typically cover places like New York and London.

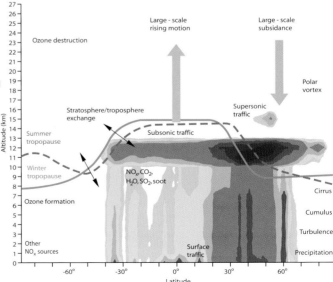

sulphur by mass. Historically all the sulphur in the fuel was thought to exhaust as SO_2, but more recently it has been found that small amounts of SO_3 and H_2SO_4 are exhausted, both of which are important in contrail formation.

Particulate matter (smoke)

Globally, natural sources like volcanoes and dust provide a significant portion of this pollution species; however, man-made sources from engine emissions can dominate in populated areas. There are growing fears that exposure to particulate matter could cause breathing disorders or cancer. It has also been suggested that the direct injection of particulate matter into the atmosphere can contribute to cirrus cloud formation.

The formation of smoke is dependent on the air/fuel ratio and pressures and temperatures within the combustor; the highest smoke production occurs at medium or high engine power. Modern combustors are designed to produce no visible smoke.

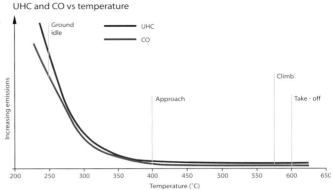

The UHC and CO emission characteristics of a jet engine combustor.
The highest emissions are at idle where low temperatures result in the incomplete combustion of the fuel.

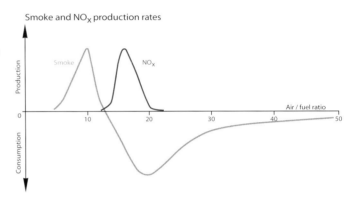

Smoke and NO_x production and consumption in the combustor varies with local air/fuel ratio values

Airport pollution and the LTO cycle

Modern gas turbine engines have come a long way since the early inefficient, noisy, visibly smoky, and malodorous engine designs. The high bypass ratio has produced a much quieter engine, and improved knowledge has enabled engineers to almost eliminate low power emissions such as UHC, CO, and smoke. Surveys carried out in the early 1990s revealed that aircraft emissions only contribute to 20 per cent of the total NO_x at terminals – the rest being a mixture of local industry and land transport.

Once it became apparent that some sort of emissions monitoring was needed at airports, a standard landing and take-off cycle was devised – the LTO cycle. This cycle is based on aircraft activity in and around airports, and, as such, takes no account of aircraft flight emissions beyond 3,000ft. From this standard cycle, ICAO (International Civil Aviation Organization) regulations have been imposed that monitor the engine performance, not taking into account any airframe factors. Engine certification is based on this cycle and the sum of the pollutants over the cycle must be below the ICAO limit.

Future trends

It is generally recognised that the influence of aircraft on the earth's temperature change is too small to detect at this time and will remain undetected for many years. This makes it impossible to verify any results or predictions at present. It is also difficult to separate the aircraft-only signal from the effect of other anthropogenic changes in ozone and carbon dioxide.

One option could be flying at higher altitudes. This might be environmentally acceptable because of reduced contrail formation if it can be conclusively shown that the chemical effects of the emissions are of minor importance and if fuel consumption is smaller than for present aircraft.

The standard LTO cycle along with the respective power settings and the duration at which the engine operates at each setting

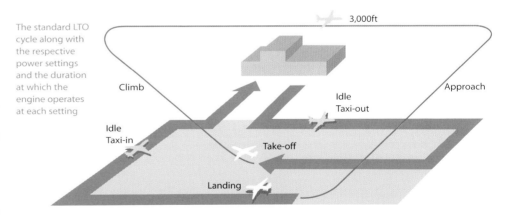

Operating mode	Power setting	Time in mode
1 Taxi/idle	7% take-off thrust	26 minutes
2 Take-off	100% std. day take-off thrust	42 seconds
3 Climb	85% take-off thrust	132 seconds
4 Approach	30% take-off thrust	4 minutes

Environmental impact is an undesirable by-product of the gas turbine.
Performance is its prime function.

performance

AMBIENT TEMPERATURE, TURBINE ENTRY TEMPERATURE, TURBINE OPERATING TEMPERATURE (AT VARIOUS STAGES), PRESSURE RISE THROUGH THE COMPRESSOR, AIRFLOW, FUEL FLOW, BYPASS RATIO, DRAG, ACCELERATION, DECELERATION: THE NUMBER OF VARYING CONDITIONS THAT INFLUENCE AN ENGINE'S PERFORMANCE IS ALMOST INCALCULABLE.

performance

Performance is the thrust or shaft power delivered for a range of given parameters:

> fuel flow

> life

> weight

> emissions

> engine diameter

> cost.

Performance engineering has two pivotal roles: first, it ensures stable engine operation throughout the operational envelope, under all steady state and transient conditions; second, it integrates component technologies so that the product attributes critical to the end user, are optimised for any given application.

Performance is critical to all phases of gas turbine design, development, and operation. It is also a significant part of what a gas turbine manufacturer sells and the operator buys.

The operating condition where the engine will spend most of its time has traditionally been chosen as the engine design point. For a long-range, civil airliner, this would be its cruise condition, typically 35,000ft, Mach 0.82 to Mach 0.85 on a standard (ISA) day. It is primarily at this operating condition that the engine performance, configuration, and component design are optimised, though the latter two are heavily influenced by more arduous flight conditions.

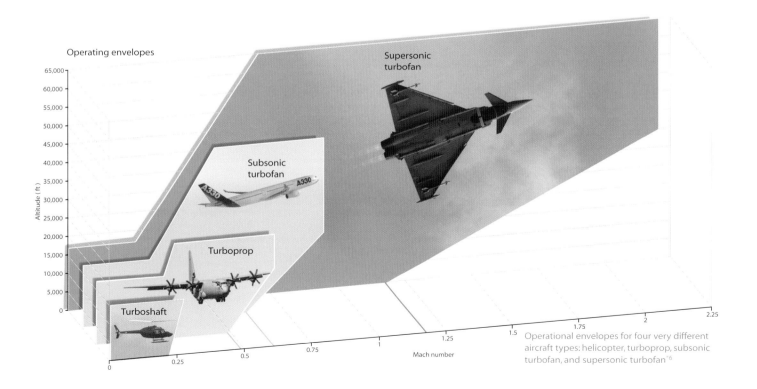

Operating envelopes

Supersonic turbofan

Subsonic turbofan

Turboprop

Turboshaft

Altitude (ft)

65,000
60,000
55,000
50,000
45,000
40,000
35,000
30,000
25,000
20,000
15,000
10,000
5,000
0

Mach number

0   0.25   0.5   0.75   1   1.25   1.5   1.75   2   2.25

Operational envelopes for four very different aircraft types: helicopter, turboprop, subsonic turbofan, and supersonic turbofan[76]

Design point performance and engine concept design

A number of design point performance parameters can be used to give an initial, or first order, comparison of the overall performance of competing concept designs (» 41):

> Specific thrust is the output thrust divided by the engine inlet mass flow; specific power is similar, based on output power. This provides a good, first order indication of the engine weight, frontal area, and volume for a given thrust.

> Specific fuel consumption (sfc) is the fuel flow rate divided by the output thrust or power. For long-range, civil aircraft engines, a low sfc is critical as the cost of fuel is typically 15 to 25 per cent of aircraft operating costs.

There are a number of gas turbine cycle parameters that have a powerful effect on sfc and specific thrust or power. For a turbojet,

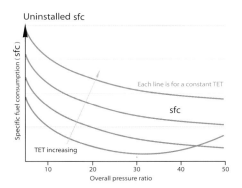

Uninstalled sfc

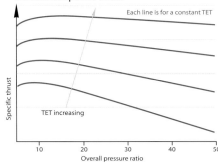

Uninstalled specific thrust

These design point diagrams show how the specific thrust and sfc of a turbojet are influenced by compressor pressure ratio and turbine entry temperature. Each point on a design point diagram represents a different engine geometry.[6]

these are compressor pressure ratio and turbine entry temperature (TET).

Specific thrust improves dramatically with turbine entry temperature, and the optimum pressure ratio is about 8:1 at low TET and 15:1 for high TET. Conversely, sfc gets worse as TET is increased but improves as pressure ratios become higher.

The concept designer must, therefore, make a compromise between achieving the best sfc or specific thrust when choosing the cycle parameters. Many other limitations must also be considered including the complexity of engine design resulting from a very high pressure ratio and the mechanical integrity limitations of going to a very high turbine entry temperature. As component efficiencies improve, so do the absolute levels of both specific thrust and sfc, but the fundamental shape of the design point diagrams does not change.

For aero engines, sfc can be considered to have two components. Thermal efficiency is the rate of addition of kinetic energy divided by the rate of fuel energy supplied, whereas propulsive efficiency is the useful power produced divided by the kinetic energy supplied.

$$sfc = \frac{3{,}600V_{flight}}{\text{Thermal efficiency*propulsive efficiency*LHV}}$$

Where V_{flight} is free stream air velocity (flight velocity), LHV is the fuel lower heating value (J/Kg.K, commonly called calorific value), and '3,600' converts from seconds to hours – sfc is measured in kilograms of fuel burnt per hour per Newton of thrust.

Propulsive efficiency can be shown to be

$$\text{Propulsive efficiency} = 2V_{flight}/(V_{flight} + V_{jet})$$

where V_{jet} is exhaust velocity from the propelling nozzle.

Hence, for a given flight speed, propulsive efficiency and sfc will both improve as jet velocity is reduced. However, the equation for thrust

$$F = W(V_{jet} - V_{flight})$$

The optimum fan pressure ratio for sfc and specific thrust reduces with bypass ratio. Specific thrust deteriorates with bypass ratio whereas sfc improves with bypass ratio – as it does with core pressure ratio. At high bypass ratios, increasing TET can improve sfc. Increasing TET always improves specific thrust.

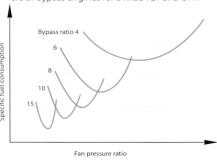

sfc of bypass engines for a fixed TET and OPR

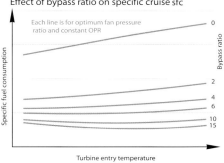

Effect of bypass ratio on specific cruise sfc

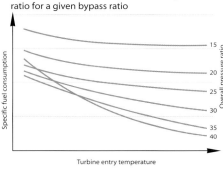

Effect on cruise sfc of temperature and pressure ratio for a given bypass ratio

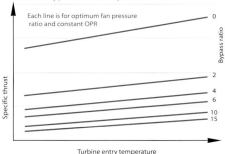

Effect of bypass ratio on specific thrust

The EJ200 in the Typhoon uses reheat to provide extra thrust

shows that as jet velocity is reduced, increasing mass flow W is the only way to maintain thrust F at the level required for the concept design point. This is the fundamental driver for the turbofan engine where the bypass provides a jet of high mass flow and low jet velocity. For civil aircraft applications, the improvement in sfc far outweighs the deterioration in specific thrust.

For turbofan engines, the bypass ratio and the fan pressure ratio are additional cycle parameters to the core overall pressure ratio and turbine entry temperature.

Once a promising design point has been selected, then the next phase in the concept design process is to freeze the engine geometry so that performance at other key operating conditions, such as sea level static take-off, can be computed. In these off-design performance calculations, geometry is fixed and the operating conditions change. In the concept design phase, design point and off-design calculations must be used iteratively so that satisfactory cruise performance can be achieved while also delivering the required take-off thrust with acceptable turbine entry temperature.

Referred parameter groups
Once an engine's geometry has been defined, then referred parameter groups become key to gaining an appreciation of how an engine (and its components) behaves at off-design and transient conditions.

For a thrust engine operating at a given flight Mach number, there would be, for example, one plot of inlet mass flow versus engine rotational speed for every combination of pressure, altitude, and inlet temperature. However, when working to first order accuracy,

this huge number of graphs can be collapsed onto a single plot by using the referred parameter groups for inlet mass flow and speed. Similarly, ignoring second order effects such as Reynolds Number, the compressor and turbine maps (>> 80) enable a single plot to be used rather than taking the raw parameters and calculating a different plot for every component inlet temperature and pressure combination. The engine working line on these maps can also be plotted in this collapsed fashion.

Off-design performance
The steady state performance of a fixed engine design varies with its current operating condition, which comprises the engine setting in terms of thrust/power level and the point within the operational envelope.

Ambient pressure and temperature vary dramatically with altitude. Under normal forward flight, total temperature and pressure at engine inlet increase from these ambient conditions. For example, at 0.85 flight Mach number, the ram effect increases inlet total pressure by a factor of about 1.6 and inlet total temperature by about 1.15.

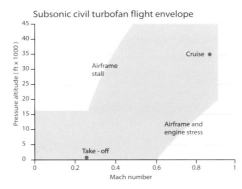

Subsonic civil turbofan flight envelope

Pressure altitude (ft x 1000) vs Mach number

- Cruise
- Airframe stall
- Airframe and engine stress
- Take-off

A typical operational envelope for a civil aero turbofan in terms of altitude and flight Mach number[6]

The key referred parameter groups for performance

Temperature at station n	T_n/T_1
Pressure at station n	P_n/P_1
Mass flow	$W\sqrt{T_1}/P_1$
Rotational speed	$N/\sqrt{T_1}$
Fuel flow	$W_F/(P_1\sqrt{T_1})$
Gross thrust	FN/P_1
Momentum drag	$FRAM/P_1$
Power	$PW/(P_1\sqrt{T_1})$
Thrust sfc	$sfc/\sqrt{T_1}$
Power sfc	sfc
Shaft acceleration	NU/P_1

Ambient pressure versus pressure altitude

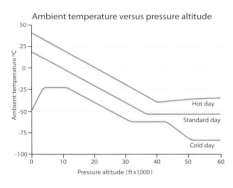

Pressure decreases with altitude – as does temperature under most circumstances[6]

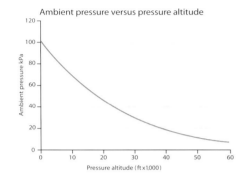

These variations in inlet conditions have a powerful impact on engine performance.

When the engine is throttled back and referred speed is reduced, then all the other referred parameter groups reduce. The effect of flight Mach number should also be noted in that, once the propelling nozzle unchokes (» 14), the referred parameters fan out from a single line. One point of particular interest is that in this low power operating regime the compressor working line is close to the stability line at lower flight Mach numbers – particularly for a fan or LP (low-pressure) compressor.

If 100 per cent referred speed could be maintained throughout the operational envelope, then all of the other referred parameter groups would be constant. Hence:

› The absolute speed varies with the square root of ram inlet temperature. It decreases, therefore, as temperature reduces with altitude, but increases with Mach number and on hot days.

› Turbine entry temperature is directly proportional to ram inlet temperature. For example, if the engine were at 35,000ft then, relative to ISA sea level static, turbine entry temperature would have decreased by a factor of 21/288.15 due to altitude, but with 0.85 Mach number it would have increased by a factor of 1.15, so the overall TET change reduction is 12.5 per cent.

› The operating point on the compressor map is unchanged throughout the operational envelope (while the final nozzle is choked).

› Gross thrust and momentum drag both decrease with altitude because ambient pressure decreases, leading to a reduction in net thrust. However, both increase with Mach number due to the ram increase of inlet pressure, P_1. Combined, these effects result in a net thrust recovery with Mach number. Due to the higher mass flow of the turbofan compared to a turbojet, the turbofan's momentum drag increases more quickly with Mach number and so net thrust recovery is worse.

Turbojet and turbofan maximum rated thrust versus flight Mach number

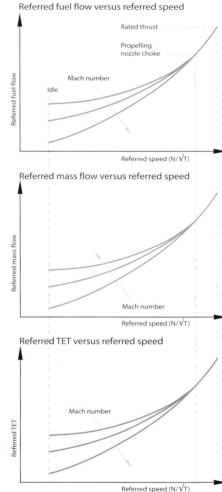

When working to first order accuracy, referred parameter groups can be used to show how turbojet performance varies throughout the operational envelope[6]

Turbojet and turbofan thrust versus mach number [6]

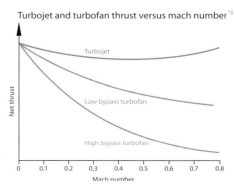

Engine ratings

Retaining engine rotational speeds, temperatures, and pressures below mechanical limits mean that, in reality, the engine cannot operate up to 100 per cent referred speed at all flight conditions. The engine control system must be set up to govern, or rate, the engine at key flight conditions so that sufficient thrust is provided but mechanical integrity limits are not exceeded.

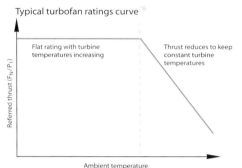

Typical turbofan ratings curve [6]

Referred thrust (F_N/P_1)

Flat rating with turbine temperatures increasing

Thrust reduces to keep constant turbine temperatures

Ambient temperature

Typically, thrust is rated against ambient temperature for take-off at sea level static conditions. Where thrust is flat rated, referred speed and referred TET are constant, but as the ambient temperature increases, the absolute speeds and TET must also increase. At a certain ambient temperature, TET usually meets its mechanical limit and the engine must then be rated to this limit with thrust falling as a result.

At the top of climb, TET and absolute speed are often not the barrier due to the much lower ram inlet temperature – TET and speed are low relative to sea level static. In this case, it may be that an upper limit to referred speed is set due to fan or compressor aerodynamic constraints.

Transient performance

Transient performance covers operating regimes where engine parameters are changing with time. Engine operation during transient manoeuvres is often referred to as handling or operability. In particular, avoiding engine instabilities such as compressor surge, where the flow in the compressor reverses violently (») 96–99), or combustor weak extinction must be balanced with achieving the engine acceleration and deceleration times required by the application.

Performance parameters vary during a slam acceleration or slam deceleration. During an engine accel in response to a step change in throttle demand, the control system increases fuel flow; this in turn increases TET and turbine output power. This higher turbine output power exceeds that required both to drive the compressor and auxiliaries and also to overcome mechanical losses. The excess power is available to accelerate the shaft with the result that airflow, pressures, and temperatures through the engine all increase. This acceleration continues until the steady state condition corresponding to the new throttle setting is reached. The opposite of this process occurs during deceleration.

It is a characteristic of gas turbine engines that the HP (high-pressure) turbine is usually choked for all operation above idle operation, and, during an accel, there is a tension

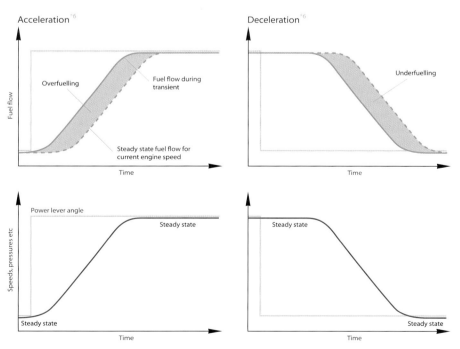

Acceleration [6]

Fuel flow

Overfuelling

Fuel flow during transient

Steady state fuel flow for current engine speed

Time

Deceleration [6]

Underfuelling

Time

Speeds, pressures etc

Power lever angle

Steady state

Steady state

Time

Steady state

Steady state

Time

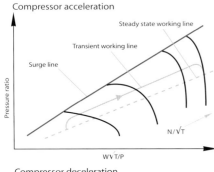

Compressor acceleration

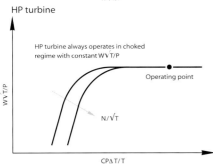

Compressor deceleration

HP turbine

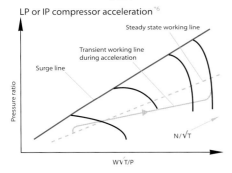

LP or IP compressor acceleration [*6]

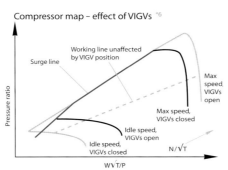

Compressor handling bleed valve [*6]

Compressor map – effect of VIGVs [*6]

between putting in enough over-fuelling to achieve the required accel time but not surging the HP compressor. When fuel is initially put in at idle, the TET rises and to keep the turbine referred mass flow ($W\sqrt{T}/P$) constant the ratio of turbine inlet pressure to flow must increase. Initially, due to the shaft inertia, the compressor speed is unable to increase – the only way the compressor can match these new turbine requirements is to go up its referred speed line towards surge. The fuel schedule must be set so that the compressor does not reach surge before the engine speed starts to increase. The turbine can then be satisfied by the transient working line running parallel to the stability line during the accel so that the increase in pressure exceeds the increase in mass flow.

For a decel, the process operates in reverse with the transient working line being lower than that for steady state. In this instance, the performance engineer must guard against combustor weak extinction due to a reduction in fuel flow. In addition, lower pressure and higher mass flow can together create adverse combustor stability conditions.

The accel and decel characteristics are different for an LP compressor or a fan from those of an HP compressor. During an accel, the LP compressor working line initially shows a small increase up its referred speed line in order to satisfy the reduced mass flow into the HP compressor. As the HP spool then accelerates, it can swallow more mass flow and the LP compressor working line is dragged below its steady state level. In a decel, the reverse is the case and so, for the LP compressor, it is during a decel that surge is an issue.

As an engine is throttled back, the steady state compressor working lines will usually head towards the stability line, which may leave insufficient margin for transient excursions such as emergency manoeuvres or accels. Two variable geometry mechanisms are commonly employed to manage this situation. First, handling bleed valves in, or downstream of, the compressor may be opened at part power. This has the effect of requiring a higher compressor mass flow thus lowering the working line to a satisfactory level. The disadvantage of bleed valves is that they increase sfc and TET at the given part power level of thrust, as well as adding cost, complexity, and weight. Second, variable inlet guide vanes (VIGVs) can be positioned in front of the compressor. Variable stator vanes (VSVs) for a number of the front stages of the compressor are sometimes also employed. These variable vanes are closed at part power, sliding the compressor map to the left. The steady state working line is essentially unchanged and so more low-power stability margin is available. Variable vanes do not have the performance penalty associated with handling bleed valves, but can be of higher cost, complexity, and weight.

Starting

Starting – the phase of operation from when the operator or pilot selects a start through to stabilisation at idle – is one of the most technically challenging aspects of gas turbine performance. For aircraft engines, restarts in flight, as well as ground starting, must be addressed.

During the dry cranking phase, the HP spool is accelerated by the starter with no fuel being metered so that sufficient pressure and mass flow can be developed in the combustor to allow it to light satisfactorily when required (» 120). In some instances, the engine may be operated at the top of crank, the maximum speed the starter can sustain, for a short time to purge fuel that may be in the gas path from previous failed starts. Fuel is then metered to the combustor and the igniters are energised. After ignition and light around, fuel flow is increased to allow the engine to accelerate to idle. The starter is disconnected from the engine during this last phase. To reduce thermal stress, the engine is usually held at idle for a time so that it can thermally soak to this condition before being accelerated further.

The design of the starter system is complex. It is critical that the impact of hot and cold days is considered. On cold days, oil viscosity will be greatly increased leading to higher engine resistance. Fuel viscosity is also higher on cold days, reducing its atomisation properties; this must be considered with respect to ignition and light off. Conversely,

on hot days, the engine fuel schedule and, therefore, acceleration power from the engine may have to be lowered due to limitations on the absolute level of TET allowed during a start for mechanical considerations. Furthermore, the assistance torque from the start system and the parasitic drag of driven accessories will vary with ambient conditions.

In the start regime, operability is a key issue. Being able to establish and maintain stable combustion at higher loadings than normal operation is very important. The other issue that must be managed is compressor rotating stall – that is the upper boundary in the sub idle regime – rather than surge. The higher the fuel schedule, the higher the transient working line on the compressor. If the fuel schedule is set too high, the HP compressor will be driven into rotating stall where its efficiency drops markedly and, for a given fuel schedule, TET will increase rapidly so the start will have to be aborted. Conversely, if the fuel

schedule is set too low, there will not be enough assistance for the engine to accelerate to idle in the required time; in the worst case, it may stop accelerating completely. It is, therefore, critical that the compressor is designed with sufficient low-speed rotating stall margin. To keep the working line as low as possible, bleed valves will be open and variable vanes closed during a start.

For manned aircraft engines, the ability to restart in flight is essential. The restart process is similar to ground starting for the starter-assisted portion of the envelope. The left hand boundary is limited by being able to achieve sufficient combustor pressure and mass flow for light off as well as having sufficient stall margin because the working line will be at its highest at low flight Mach number.

In the windmilling portion of the envelope, the starter is not employed as the ram effect of the higher flight Mach number causes the

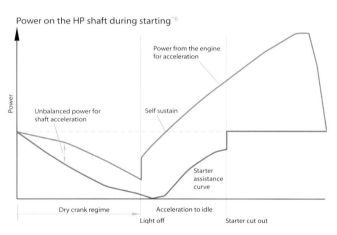

Power on the HP shaft during starting [6]

A typical assistance/resistance curve during a start. The starter input power is plotted as a negative value so that the distance between the two lines represents the net torque available for shaft acceleration.

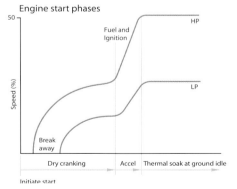

Engine start phases

The phases of a start for a two-spool turbojet or turbofan [6]

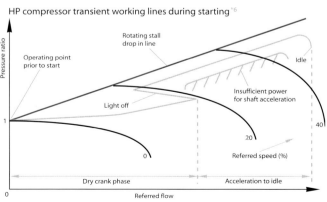

HP compressor transient working lines during starting [6]

Typical instrumentation for key sea-level, steady-state performance tests[*6]

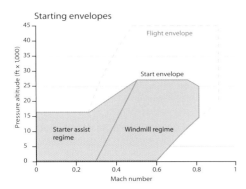

Starting envelopes

The International Standard Atmosphere, or ISA, was established by the International Civil Aviation Organization as a uniform reference for temperature and pressure.

The standard atmosphere was derived from the average conditions for all latitudes, seasons, and altitudes. The properties of a standard day are related to sea level at 45 degrees latitude with zero humidity. The standard temperature is 15°C and the standard barometric pressure is 101.325kPa (1,013.25 millibars).

Heavily-instrumented compressor test rig

engine to windmill at relatively higher rotational speed and also provides sufficient pressure and mass flow in the combustor for ignition and light off. The right hand side of this portion of the envelope is limited by combustor stability; if the gas velocities are too high in the combustor, a flame cannot stabilise.

Engine performance testing

Engine performance testing is a critical part of performance technology. During a development programme for a new engine type, an immense amount of effort is spent validating both the predictions of steady state performance throughout the flight envelope, and also transient performance and starting

predictions. Similar effort goes into ensuring the engine performance condition is right for other major integrity tests such as bird ingestion, thermal paint, or the 150-hour endurance test.

After service entry, performance pass-off testing of each individual production engine is common practice, ensuring that it meets key acceptance criteria. With the relentless drive for lower cost of ownership, more focus is being placed on performance analysis of on-wing data from engines in service.

From a performance perspective, the ideal test facility is outdoors so that the engine environment is as close to the free field case as possible. However, it is surprisingly difficult to find outdoor locations that do not have noise restrictions, that do have suitable climatic conditions to allow high initialisation, and are not so remote that the logistics of operating them become prohibitive

In most countries, therefore, indoor test facilities are used. For a given engine condition, the measured thrust on an indoor bed may be up to five per cent less than that measured on a free field facility. This is due to the inlet momentum of the airflow in the test bed (the air is not static) and the unrepresentative static pressure field around the engine and cradle caused by the velocity

of air within the test bed passing around the engine. Consequently, an indoor thrust facility must be meticulously calibrated against an outdoor facility. This is done by running performance tests using the same engine, usually in an A-B-A sequence of back-to-back tests between the two test facilities.

For transient tests, faster response instrumentation must be used so that measurements can be taken at up to 100 scans per second without the instrumentation system introducing unacceptable delay or lag.

After engine data has been recorded, a test bed analysis programme is run to calculate a range of derived parameters. These calculations include

> Applying known calibrations to go from raw signal output to engineering units such as pressure transducer milliVolts (mV) to pressure (kPa), thermocouple mV to temperature, fuel meter frequency to fuel flow in litres/s, or air meter pressures to inlet airflow in kg/s.

> Extensive error checking of the measurements.

> Where a number of pressure or temperature rakes have been used at a station, they must be suitably averaged.

> Working out parameters such as sfc, core air mass flow on a turbofan, and TET.

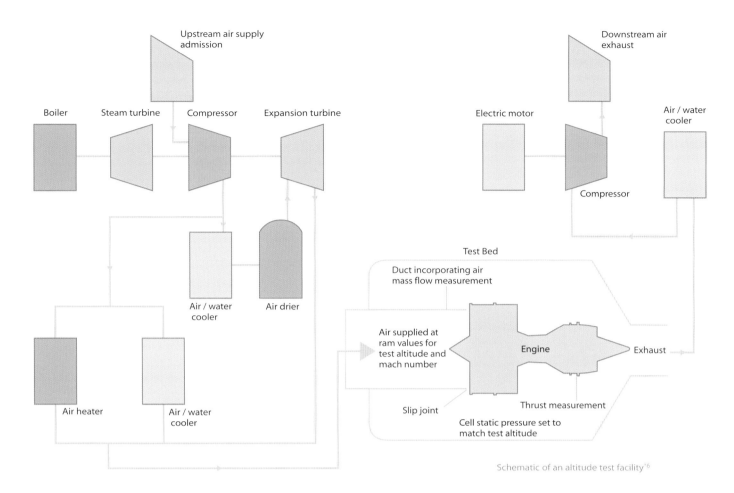

Upstream air supply admission

Boiler Steam turbine Compressor Expansion turbine

Air / water cooler

Air drier

Air heater

Air / water cooler

Downstream air exhaust

Electric motor

Compressor

Air / water cooler

Test Bed

Duct incorporating air mass flow measurement

Air supplied at ram values for test altitude and mach number

Engine

Exhaust

Slip joint

Thrust measurement

Cell static pressure set to match test altitude

Schematic of an altitude test facility[6]

》 All parameters are referred to standard day conditions of 15°C (288.15°K) and 101.325kPa using the referred parameter groups listed earlier (》 77). This enables test data run at one set of ambient conditions to be compared directly to data collected on a different day.

》 The evaluated component and engine performance levels (for example, efficiency, flow, and speed) are compared to prediction, often in an automated fashion with the steady state model being coupled to the test data analysis code.

Having made these calculations, it is critical to understand the reasons for the differences between pre-test prediction and the analysed test data.

Engines can also be tested in an altitude test facility, which reproduces the inlet ram total pressure and the temperature required for the altitude and Mach number combination under test, as well as the exit static pressure consistent with the given altitude. Considering that the mass flow being conditioned may be up to 500kg/s, it will be apparent that these plants require huge capital investment

and are very expensive to operate. However, they do allow a heavily instrumented engine, including a direct measurement of thrust, to be exercised throughout the flight envelope

Alternatively, a flying test bed may be used to measure in-flight performance. This provides better simulation of effects such as engine geometry changes with in-flight loads and pressure profiles at entry to the engine due to the intake. On the other hand, it has limited instrumentation capability and, most importantly, it is not possible to measure thrust directly.

A new aircraft and engine combination will go through an exhaustive flight test programme. During this programme, compliance testing allows the airframer and engine manufacturer to decide whether the engine has met its cruise performance guarantees.

The A340 flying test bed with three of its original engines and one Trent 900 during its development programme for the A380

Civil aircraft engines

For engines on long-range, civil aircraft sfc is absolutely critical. One per cent of cruise sfc can be worth up to $150,000 per year on a four-engined aircraft. Gas turbine engine companies will go to great extremes to improve sfc by even a tenth of a percentage point, and the level of investment in technology to improve sfc over the decades has been immense. Improvements in materials, manufacturing, cooling, and coatings technologies have allowed dramatic improvements in TET without huge increases in cooling airflows. An inexorable improvement in component efficiencies has had a very powerful effect. These efficiencies are the result of a range of activities from empirical rig testing through to the application of advanced CFD modelling. There has also been a steady increase in the cycle parameters of overall pressure ratio and bypass ratio.

Another peculiarity of civil aircraft engine performance worthy of note is the impact of the engine failure case during take-off on the required engine thrust. For civil aircraft, the maximum weight that can be carried is typically limited by runway length and the need to consider a possible engine failure at any time during take-off. At low speed, if an engine were to fail, the aircraft must be able to stop within the runway length. At higher speeds, when it would not be possible to stop, the aircraft must be able to continue its take-off with a failed engine. Therefore, compared to engines on a four-engined aircraft, an engine for a twin-engined aircraft must have a far greater thrust capability beyond that required for normal operation to cater for the failure case.

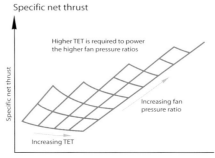

Specific net thrust

Higher TET is required to power the higher fan pressure ratios

Increasing fan pressure ratio

Increasing TET

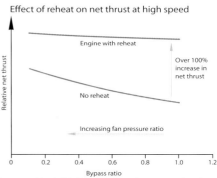

Effect of reheat on net thrust at high speed

Engine with reheat

Over 100% increase in net thrust

No reheat

Increasing fan pressure ratio

The thrust available from various choices of specific thrust at Mach 2.0. Negative net thrusts result at lower specific thrusts even at levels much higher than civil engines.

Another key performance issue for multi-engined aircraft is that they must be able to continue to fly safely with a failed engine. The twin-engined aircraft must prove it can continue with just one engine (fifty per cent power) and still maintain a satisfactory altitude to avoid high ground. This results in the definition of a suitable thrust rating. There are no specific rules for four-engined aircraft; agreements are made between the airframer and their authority.

Military aircraft engines

Military engines are typically required to offer far greater agility and top speed capability – speeds of Mach 2.0 to 3.0 are not uncommon. As speed rises, so the relation between gross thrust and momentum drag changes, and an engine of high specific thrust (high thrust at an air mass flow) becomes desirable.

Another consideration is the drag of the engine, which is related to the physical size of the fan. This again drives designs to higher specific thrust. To achieve high specific thrust, the engine requires a high final nozzle pressure ratio, and, due to limitations on operating temperature, this necessitates a low bypass ratio (typically less than one) and a high fan, or LP compressor, pressure ratio (up to more than 5:1). Plain turbojets may also be used, but this usually results in too great an sfc penalty at lower speeds resulting in a restricted operating range.

A further consideration in military engine cycle choice, due to high Mach number operation, is the overall compressor pressure ratio. At Mach 2.2, the engine inlet temperature is over 150°C, and, since compressor delivery temperature is proportional to engine entry temperature (at a pressure ratio), this becomes a mechanical design limitation. Overall pressure ratios for such military engines are limited below 30:1 as opposed to figures approaching 50:1 for large civil engines.

Another means of increasing thrust at high flight speed is reheat, otherwise known as afterburning. This is achieved by adding fuel downstream of the point where core gases and bypass air mix to increase, and potentially double, the temperature at the nozzle exit.

Since gross thrust is

$$F_G = W_9 V_9 + A_9 (P_9 - P_0)$$

this can have a significant effect on thrust without significantly changing the operation of the turbomachinery although a variable geometry final nozzle is required. W_9 (mass flow at exit) will be near constant, as will P_9 (nozzle exit pressure) and P_0 (ambient or

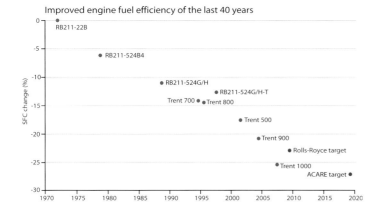

Improved engine fuel efficiency of the last 40 years

RB211-22B

RB211-524B4

RB211-524G/H

RB211-524G/H-T

Trent 700 ● ● Trent 800

Trent 500

Trent 900

Rolls-Royce target

Trent 1000

ACARE target ●

Levels of sfc have improved dramatically over the last 50 years, and will continue to improve in the future

engine inlet pressure); however, V_9 (jet velocity) and A_9 (nozzle area) are proportional to $\sqrt{T_9}$ (jet temperature) and A_9 is proportional to $W\sqrt{T}/P$.

Another area peculiar to military applications is short take-off and vertical landing (STOVL). This requires the use of thrust vectoring nozzles to change the direction of thrust from the horizontal required for forward flight to vertical for hovering. Particular performance issues raised by this include the potential for hot exhaust gases to enter the intake – potentially leading to surge or stall – and the need for the engine to decelerate rapidly on landing. This requirement for a rapid deceleration is because, on a constant velocity vertical descent, engine thrust equals aircraft weight, and so when the aircraft lands it will tend to bounce and then rise at a similar rate. Thrust has to be quickly reduced to prevent this, leading to potential surge of the low-pressure compressor.

Both missile firing and the operation of the aircraft at a high angle of attack for rapid manoeuvring can potentially lead to compressor surge: the former due to the unsteady and uneven temperature profile entering the engine; the latter because of pressure profiles caused by operating the intake at high incidence.

Industrial applications

There are two major industrial applications of gas turbines: the first is electricity generation; the second, compressing natural gas or pumping oil along pipelines that can be thousands of miles long from the well head to the end user.

In modern two- or three-shaft industrial aero-derivatives, the LP turbine system is modified or replaced and the propelling nozzle removed. The available expansion is used to provide output shaft power, via extra LP turbine stages, instead of jet thrust. Other changes to the configuration are an industrial style inlet and exhaust, usually a low emissions combustor, and some changes to materials and coatings to cater both for longer operating hours with fewer cycles; and for the corrosive effects of offshore air or even diesel fuel.

A free power turbine engine can be used for both power generation and in the oil and gas industry. The gas generator spool can be throttled up and down with its referred parameters behaving just like an aero turbojet, but obviously always at static conditions. However, for a given gas generator operating point, the free power turbine can have a range of power and speed combinations. For a given gas generator speed, the output power varies due to the power turbine efficiency changing with power turbine speed.

A compressor inside a natural gas pipeline will demand many different output power and speed combinations. However, if the

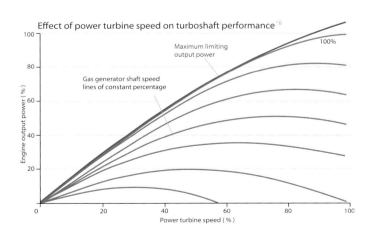

Effect of power turbine speed on turboshaft performance [6]

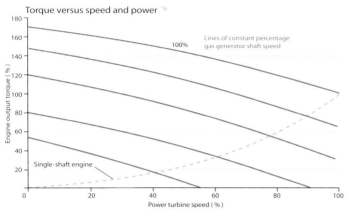

Torque versus speed and power [6]

Lines of constant percentage gas generator shaft speed

100%

Single-shaft engine

Output torque versus gas generator shaft speed and power turbine speed for an industrial engine

engine is being used for power generation, then the output speed must be held constant at all power levels as the generator must be run at a synchronous speed to maintain a steady 50Hz or 60Hz generation. The engine, therefore, will always run vertically up and down the 100 per cent power turbine speed line, as the gas generator operating speed changes.

Industrial specific, that is non-aero-derivative, engines employ a single-shaft configuration, which is only suitable for electricity generation. For the free power turbine engine, torque rises as power turbine speed is reduced for a given gas generator speed. This is because the gas generator speed is independent and it can still be at 100 per cent speed delivering maximum mass flow, pressure, and temperature to the power turbine, even when the power turbine is at low speed. However, for the single-spool configuration, output torque falls with output speed, due to the reduction of output power.

These low levels of torque at low output speed are neither consistent with the demands of the oil and gas industry nor those of other mechanical drive applications such as marine propulsion. Some aero-derivatives also drive an LP compressor from the LP turbine as well as the output load. Where the LP compressor pressure ratio is low, the output power and speed characteristic remains acceptable for oil and gas applications.

Another peculiarity of industrial engines is the opportunity to apply yet more complex cycles and configurations; for example, a combined cycle plant where the waste heat in the exhaust is used in a heat recovery steam generator to raise steam. This is used to drive a steam turbine that in turn drives a second generator. This can take the thermal efficiency of the plant (defined as useful power out divided by fuel energy in) from about 40 per cent to approaching 60 per cent. However, such a plant has a

much higher capital cost and is less flexible in certain areas – for example, length of start time.

In combined heat and power applications, the steam can be used for space heating or the exhaust heat can be used directly in processes such as paper or cement production. Here thermal efficiencies of over 80 per cent are achievable.

Marine applications

The power required to propel a ship increases with ship speed as a cube law. This means that if the gas turbine is driving a water jet or propeller, then it must be of free power turbine configuration for the reasons described above. Considerable attention has been given to integrated electric propulsion where the engine drives a generator. The power is then delivered to a busbar from which can be drawn the ship's load required for passenger appliances and the power required for propulsion via electric motors. While in this instance a single-spool machine could theoretically be used, it is likely that only free power turbine engines, which have the flexibility to operate in all configurations, will be adapted for operation in the corrosive marine environment.

Naval vessels spend the majority of their time cruising at 10 to 15 knots, but in an emergency need to operate at over 30 knots. Due to the nature of the cube law, discussed above, the engine will be operating at part power at cruise. Fuel consumption at part

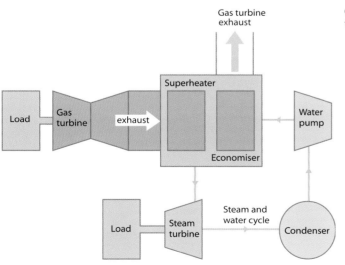

Combined cycle gas turbine plant [6]

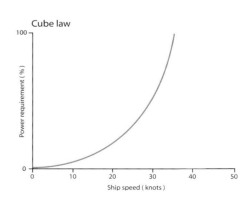

The WR-21 intercooled, recuperated marine gas turbine

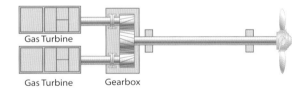

Intercooled recuperated engine fuel saving

Specific fuel consumption

ICR

Conventional

Power (%)

0 10 20 30 40 50 60 70 80 90 100

power is, therefore, very important in these applications. This is a challenge for marine engines because sfc increases significantly as a conventional gas turbine is throttled back.

Historically, a number of multi-engine configurations have been employed to

overcome this characteristic such as the CODAG (combined diesel and gas turbine) layout. Here, a diesel engine is used to provide propulsive power at low ship speed while at high speeds the gas turbine is started, providing the relatively large additional power requirement dictated by the cube law.

Other configurations have included CODOG (diesel or gas turbine) COGAG (combined gas turbines) and COGOG (small gas turbine or large gas turbine).

At the time of writing, two significant, new, marine gas turbine engines are the MT30 and the WR-21. The MT30 uses a four-stage free power turbine to maintain efficiency down to 25MW; the WR-21 is a 26MW intercooled and recuperated (ICR) engine where the heat exchangers and variable power turbine nozzle guide vanes provide a very flat sfc curve to suit naval applications, without the need for an additional small cruise engine.

Marine gas turbine powerplant configuration (COGAG / COGOG)

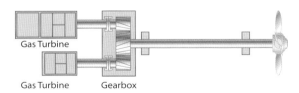

Gas Turbine

Gas Turbine   Gearbox

Gas Turbine

Gas Turbine   Gearbox

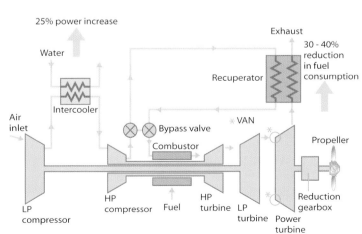

25% power increase

Water

Exhaust

30 - 40% reduction in fuel consumption

Recuperator

Air inlet

Intercooler

VAN

Bypass valve

Combustor

Propeller

LP compressor

HP compressor   Fuel   HP turbine   LP turbine   Power turbine

Reduction gearbox

Schematic of a intercooled, recuperated gas turbine

After the whole engine design, the component definition.
Beginning at the front of the engine with fans and compressors.

fans and compressors

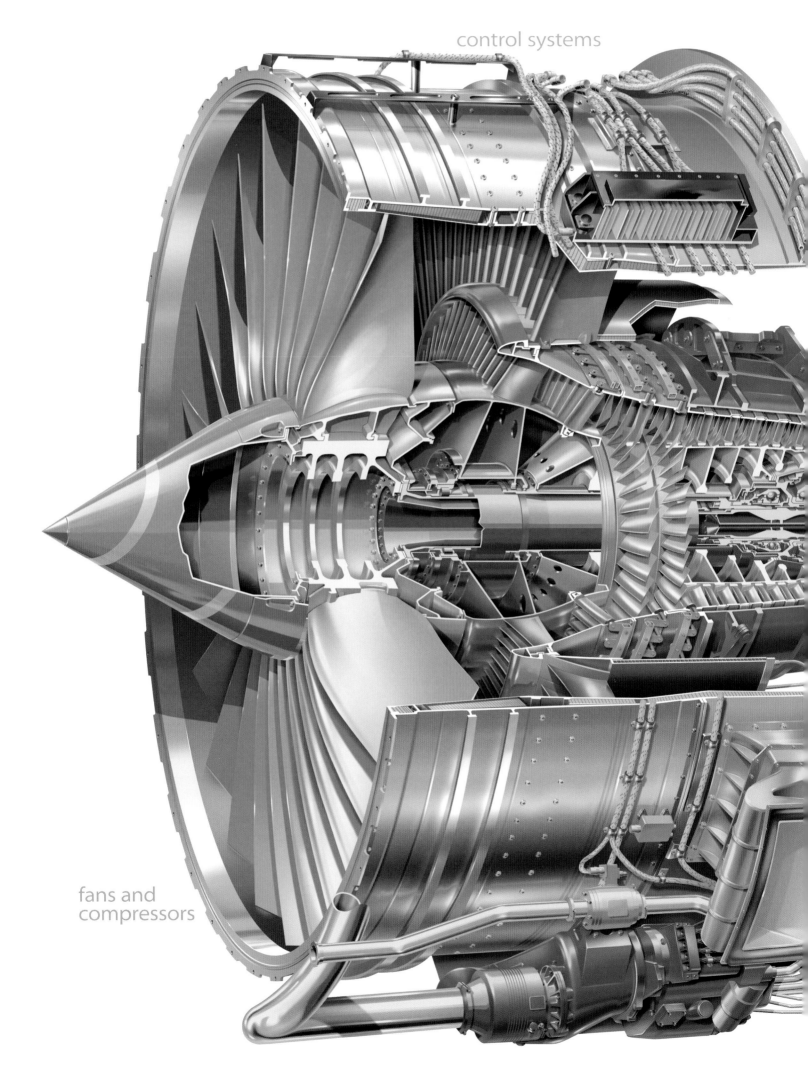

control systems

fans and
compressors

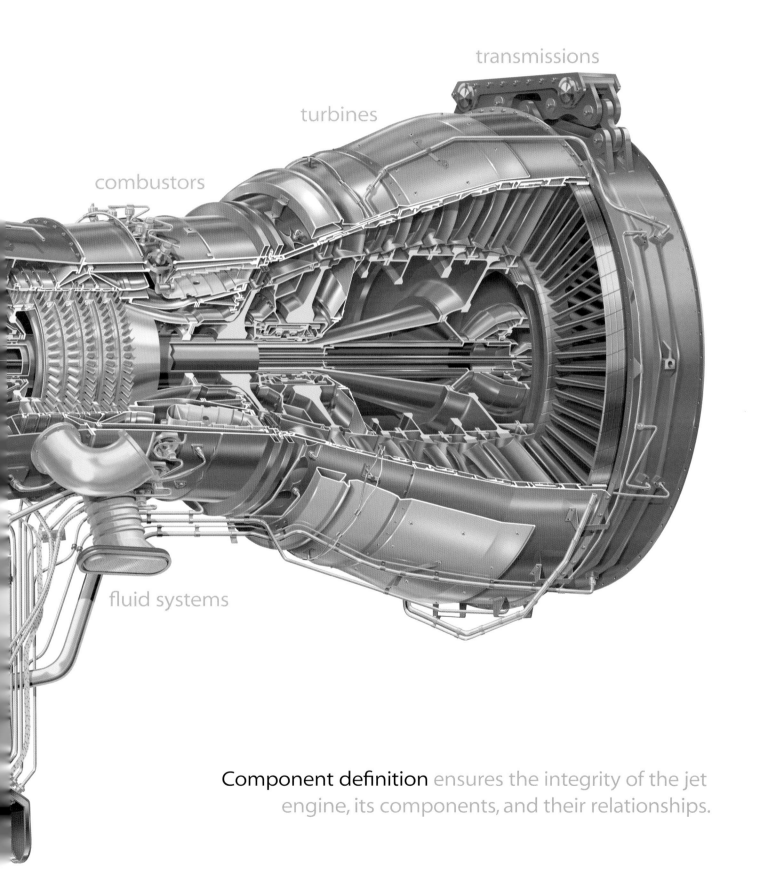

transmissions

turbines

combustors

fluid systems

Component definition ensures the integrity of the jet engine, its components, and their relationships.

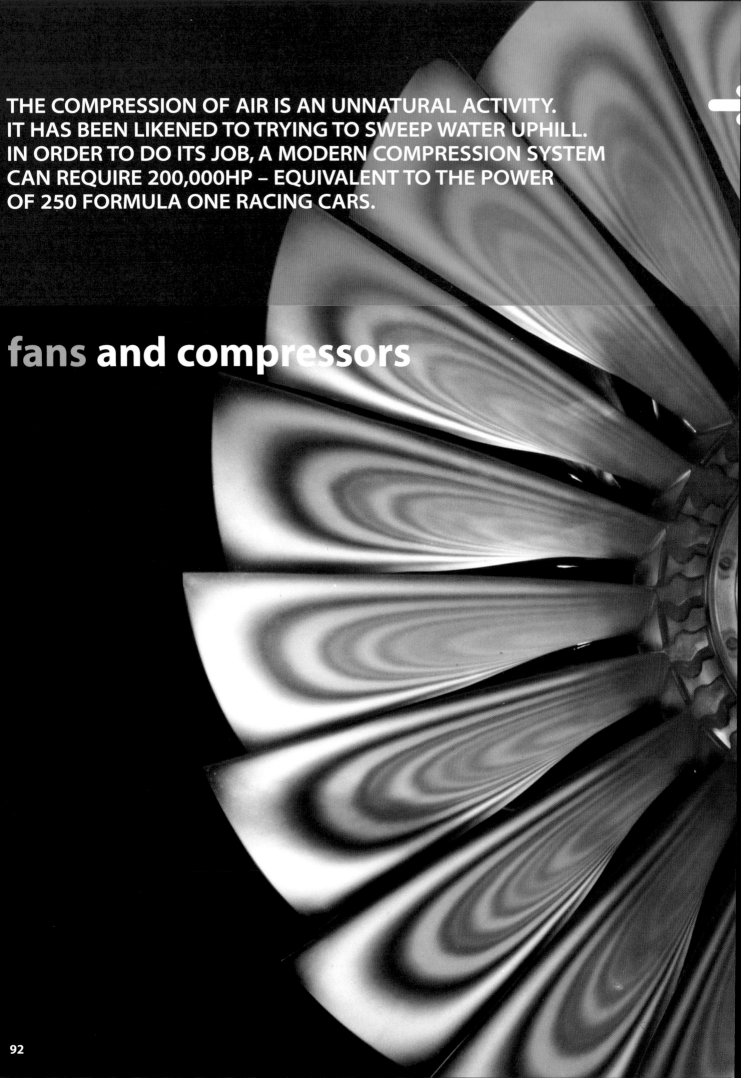

THE COMPRESSION OF AIR IS AN UNNATURAL ACTIVITY.
IT HAS BEEN LIKENED TO TRYING TO SWEEP WATER UPHILL.
IN ORDER TO DO ITS JOB, A MODERN COMPRESSION SYSTEM
CAN REQUIRE 200,000HP – EQUIVALENT TO THE POWER
OF 250 FORMULA ONE RACING CARS.

fans and compressors

A compressor is a device that raises the pressure of the working fluid passing through it – in this case, air. A fan is a large, low-pressure, compressor found at the front of most modern aero engines.

For a modern large civil engine:

〉 the fan passes over one tonne of airflow per second; this flow produces around 75 per cent of the engine thrust

〉 overall compression system pressure ratios are now approaching 50:1, and compressor exit temperatures can be over 700°C.

The design of the compression system is a complex inter-disciplinary task. Aerodynamics, noise, mechanics, manufacturing, and cost are all modelled during this process. The optimum configuration for each application is determined by performing a series of trade studies that consider all the leading attributes and requirements of the system, including life-cycle cost, weight, performance, and noise.

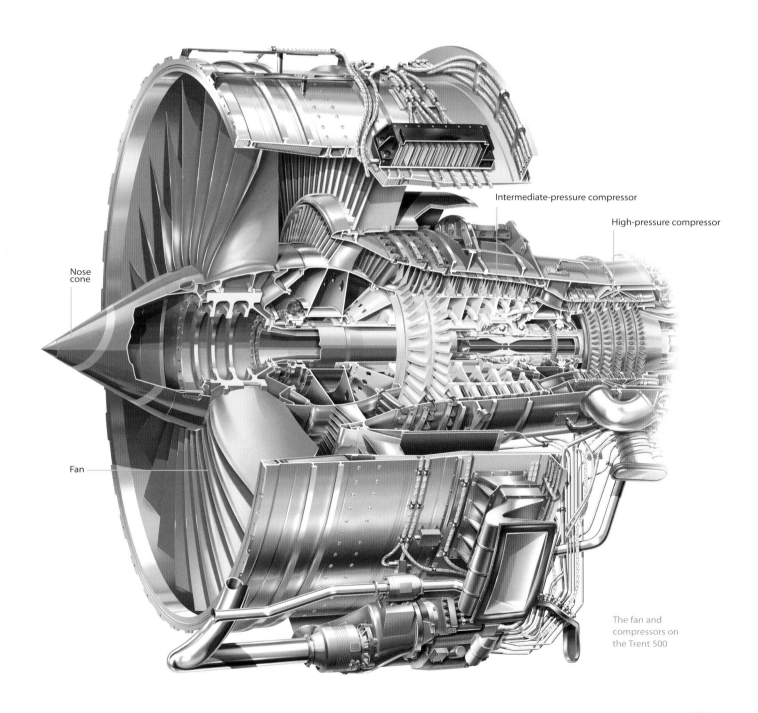

Nose cone

Fan

Intermediate-pressure compressor

High-pressure compressor

The fan and compressors on the Trent 500

Compressor configurations

For gas turbine applications, there are two types of compressor:

> axial

> centrifugal.

These two types can also be used in combination to form an axi-centrifugal (Axi-CF) compressor.

While early jet engines used centrifugal compressors, modern jet engine compression systems almost exclusively use axial compressors because a much higher compression efficiency is possible with this configuration.

Centrifugal or Axi-CF compression systems are still used for very small compressor applications as axial compressors tend not to work efficiently when the exit blade height falls below one centimetre. The centrifugal and Axi-CF systems are, therefore, more common for very small turbofans and turboshaft engines.

Compressor aerodynamics

Principles of axial compressor operation

An axial compressor consists of one or more rotor assemblies that carry rotor blades of aerofoil cross-section. The rotor is located by bearings, which are supported by the casing structure. The casing incorporates stator vanes also of aerofoil cross-section, which are axially aligned behind the rotor blades. Each rotor and downstream stator row form a stage.

The compressor rotor is driven by the turbine, via a connecting shaft. It is rotated at high speed by the turbine causing air to be continuously induced into the compressor. The pressure rise results from the energy imparted to the air by the rotor. The air is then passed through the downstream stator, where swirl is removed and a rise in static pressure achieved. The rise in the stage total pressure is proportional to the change in tangential or whirl velocity across each stage.

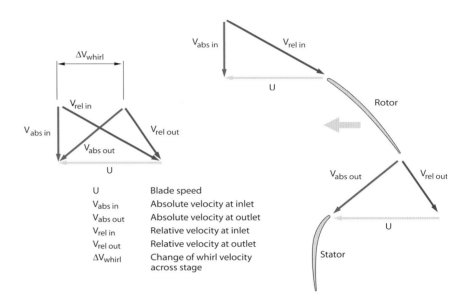

U	Blade speed
$V_{abs\ in}$	Absolute velocity at inlet
$V_{abs\ out}$	Absolute velocity at outlet
$V_{rel\ in}$	Relative velocity at inlet
$V_{rel\ out}$	Relative velocity at outlet
ΔV_{whirl}	Change of whirl velocity across stage

Compressor velocity triangles: these schematically show the passage of air through the rotors and stators

Pressure and temperature rise

As the air passes through each stage, the air pressure and temperature increase progressively. The last stator in the core or bypass stream removes all circumferential velocity, or swirl, from the air. The core air passes into the combustor pre-diffuser, before entering the combustion system.

From the front to the rear of the compressor, there is a gradual reduction of annulus area to maintain the axial velocity at a near constant level. This is usually achieved by a rising hub line or falling casing line.

For core compressors, the ratio of total pressure across each stage is in the range 1.3 –1.4. The reason for the small pressure increase through each stage is that the rate of deceleration, or diffusion, of the airflow through each of the blades and vanes must be limited to avoid losses due to flow separation and subsequent blade stall. Although the pressure ratio of each stage is relatively small, there is an overall increase in pressure across every stage. The ability to design multi-stage, axial compressors

with controlled air velocities and attached flow minimises losses and results in high efficiency and low fuel consumption.

Compressor characteristics

Under engine steady state operating conditions, the compressor will operate on the working line. However, during transient operations like acceleration, the compressor operating point can move above the working line. It is therefore vital that enough stable operating margin (stability margin) exists above the working line for any transient operation. The limit of stable operation is usually governed by the stability line.

Each stage within a multi-stage compressor possesses its own aerodynamic performance and handling characteristics – known as stage characteristics – that are subtly different from those of its neighbouring stages. Accurate matching of the stages is of crucial importance to achieving low losses and adequate operating range for off-design operation. The front stages tend to control the low speed stability margin; the rear stages, the high-speed stability margin.

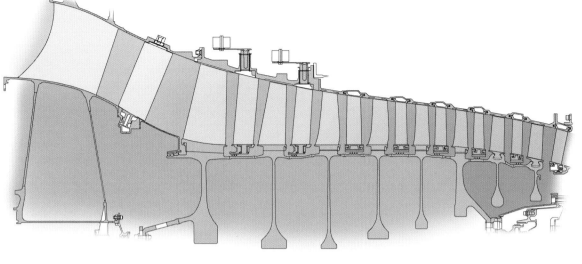

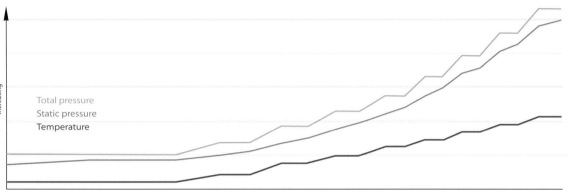

Airflow through an IP compressor – the pressure and temperature rise in the rotor, because energy is imparted to the flow; the static pressure rises in the stator due to the increasing passage flow area as the swirl is removed

Increasing pressure and temperature through compressors

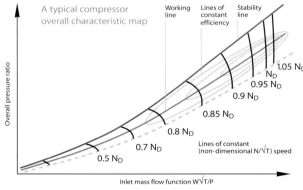

The compressor map shows the sum of all stages at any operating condition. The points can be broken down to show each stage individually. There is a significant difference between the first and last stage characteristics on a large axial compressor.

At higher operating speeds, if the operating conditions imposed upon the compressor force operation beyond the limits of the stability line, the rear stages will become overloaded, and an instantaneous breakdown of the airflow through the compressor occurs, leading to surge. During the surge event, the inlet mass flow varies with time, as the compressor flow oscillates between stalled and unstalled flow at a frequency typically around 5Hz. Due to the loss of pressure rise capability across the compressor stages, the high-pressure air in the combustion system may be expelled forward through the compressor (negative flow direction) resulting in a loss of engine thrust. This 'deep' surge produces a loud bang, and it is possible for combustion gas to come forward through the compressor inlet. Surge can also take a milder form, producing an audible 'burble', and a small fluctuation in inlet mass flow rate.

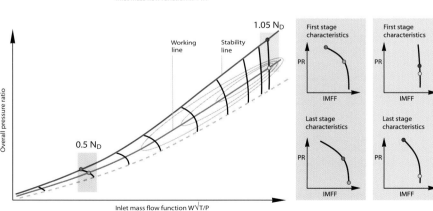

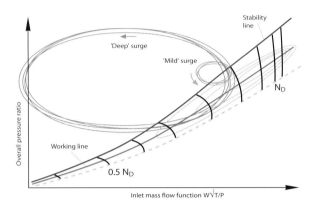

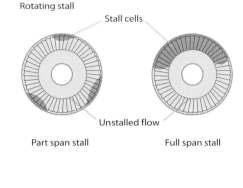

The deep and mild surge cycles illustrated on the compressor map. The compressor can go through several cycles before recovery of stability

Rotating stall cells in an axial compressor. Span refers to the radial height, or length, of the compressor blades.

At lower speeds, if the operating point is moved beyond the stability line, the front stages of the compressor may go into rotating stall. Onset of stall can be either progressive or abrupt, and is dependent on stall cell structure – part span or full span respectively. Rotating stall is non-axisymmetric, and gives rise to a circumferentially non-uniform flow, which rotates around the annulus at 20 to 50 per cent of rotor rotational speed, and in the same direction. Rotating stall frequencies are typically of the order of 100Hz.

Compressor stability at higher speeds can also be affected by the presence of flutter – a self-excited oscillation that occurs close to the natural frequency of compressor aerofoils, and results from unsteady aerodynamic loading. Flutter prediction is very complex.

Rotating stall, surge, and flutter cause blade vibration, and can induce rapid aerofoil failure and subsequent destruction of the compressor.

Compressor handling features

The more the pressure ratio of a compressor is increased, the more difficult it becomes to ensure that it will operate efficiently and in a stable manner over the full speed range. This is because the requirement for the ratio of inlet area to exit area at the high-speed operating point results in an inlet area that becomes progressively larger relative to the exit area. As the compressor speed and hence pressure ratio is reduced, the axial velocity of the inlet air in the front stages becomes low relative to the blade speed; this increases the

incidence of the air onto the blades to the point where aerodynamic stall occurs; lift is lost from the aerofoil, and the compressor flow breaks down. Where high-pressure ratios are required from a single compressor module, this problem can be overcome by introducing variable inlet guide vanes (VIGVs) and variable stator vanes (VSVs) to the front stages of the system. By closing these vanes at low speed, the incidence of the airflow onto the front stage rotor blades is reduced to angles they can tolerate.

The variable vane is of aerofoil cross-section, with an integral spindle to allow rotation, or variation of stagger. The vane is mounted in bushes in the casing or inner shroud ring and has a lever fitted to its outer end. The variable vanes' levers are all connected to the unison ring via spherical bearings, so, when the unison ring is rotated, the vanes all re-stagger together.

Alternatively, it is possible to use either bleed off-take or casing treatments to aid

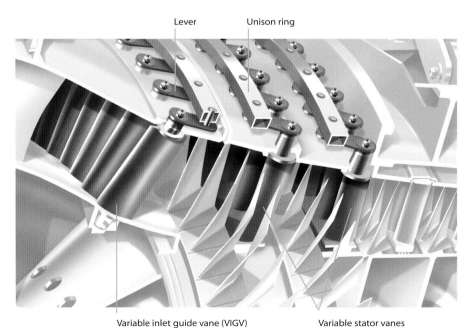

The VIGVs and two VSV stages of the IP compressor of the Trent 500

98

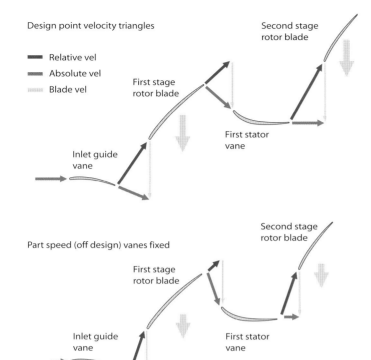

Design point velocity triangles

- ▬ Relative vel
- ▬ Absolute vel
- ▥ Blade vel

Second stage rotor blade

First stage rotor blade

First stator vane

Inlet guide vane

Part speed (off design) vanes fixed

Second stage rotor blade

First stage rotor blade

First stator vane

Inlet guide vane

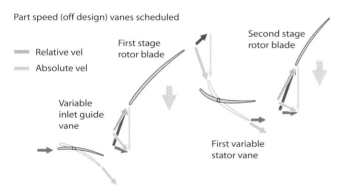

Part speed (off design) vanes scheduled

- ▥ Relative vel
- ▥ Absolute vel

First stage rotor blade

Second stage rotor blade

Variable inlet guide vane

First variable stator vane

Design point and part-speed VIGV and VSV operation

Cutaway through a three-shaft IP compressor showing VIGVs, vane levers, and unison ring

Compressor bleed air

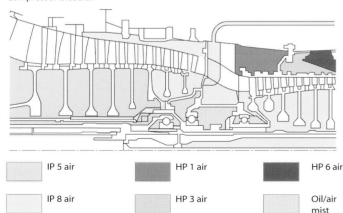

▢ IP 5 air	▢ HP 1 air	▢ HP 6 air
▢ IP 8 air	▢ HP 3 air	▢ Oil/air mist

The compressor bleed off-takes provide sealing and cooling air throughout the engine

part-speed operation. The incorporation of interstage bleeds removes a proportion of the air entering the compressor at an intermediate stage and dumps the bled air into the bypass flow. While this method corrects the axial velocity through the preceding stages, energy is wasted through the work done to compress air that is then not used for combustion, and so the use of variable stators is preferred. Bleed air can also be removed between compressor modules in order to improve engine handling. Casing treatment is another technique for improving part-speed operation. Casing treatment can be fitted to the front stage rotors to improve their stalling range, thereby improving the part-speed surge margin and engine operation. These can be slots or circumferential grooves.

Principles of centrifugal compressor operation

The impeller is rotated at high speed by the turbine, and air is continuously induced into the centre of the impeller. Centrifugal action causes it to flow radially outwards along the vanes to the impeller tip. This accelerates the air, and also causes a rise in pressure. To maximise compressor efficiency and operability, the engine intake duct may contain vanes that provide an initial swirl to the air entering the compressor impeller.

The air, on leaving the impeller, passes into the radial diffuser section where the passages form divergent nozzles that convert most of the kinetic energy into pressure energy. In practice, it is usual to design the compressor so that about half of the pressure rise occurs in the impeller and half in the diffuser. Upon leaving the radial diffuser, the air is collected in the exit system where it is further diffused.

To maximise the airflow and pressure rise through the compressor, high impeller rotational speed is required. Therefore, impellers are designed to operate at tip speeds of up to 670 metres per second (2,200 feet per second). By operating at such high tip speeds, the air velocity from the impeller is significantly increased so that greater energy is available for conversion to pressure.

To maintain the efficiency of the compressor, it is necessary to prevent excessive air leakage between the impeller and the casing. This is achieved by keeping their clearances as small as possible.

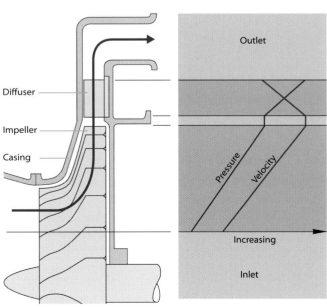

Centrifugal compressor

The Turbomeca centrifugal compressor in the RTM322[*2]

Pressure and velocity changes in a centrifugal compressor

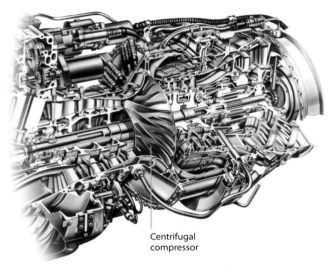

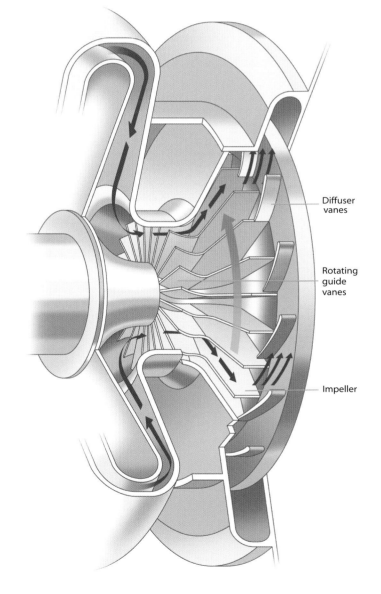

Airflow through a centrifugal compressor

Compressor subsystem description

A compressor can comprise multiple stages, but within the compression system itself there can also be multiple compressors. This multi-spool configuration consists of two or more compressors, each driven by their own turbine at an optimum speed thus achieving a higher overall pressure ratio and greater operating flexibility.

Although a multi-spool compressor can be used for a pure turbojet engine, it is more suitable for the bypass engine, where there is a splitter downstream of the first compressor module – the splitter is a mechanism to separate the intake air into core and bypass streams. The first compressor (the fan) works on the total engine flow, while the core compressor(s) work only on the core flow. For high bypass ratios, the total engine flow is significantly larger than the core compressor flow.

Some compression systems may also have axial stages in the core section driven by the fan shaft. These are commonly referred to as booster stages and are used on two-shaft engines to supercharge the core airflow into the HP compressor.

The nose cone

The nose cone provides the inner annulus profile in front of the fan for smooth airflow into the fan blades' roots, and must withstand bird impact, erosion, and the build-up of ice. To meet this functionality, the cone is made from glass fibre, laid and curved so that the cone achieves maximum strength. The thickness of the cone is based on the experience of bird impact tests from previous engines; the angle of the cone is optimised for both bird impact and ice shedding behaviour, while delivering satisfactory airflow into the fan. The nose cone also has a rubber tip at the front to dislodge any ice accretion. There is also a seal bonded to fit beneath the fairing, and a polyurethane coating for protection against erosion. The nose cone is generally secured to the fan module by a bolted flange and internal spigot arrangement.

The cone thickness on more recent engines has been modified to meet the heavier bird certification requirements and a double ply lay-up technique is now widely used.

The nose cone with rubber tip to prevent and remove ice build-up

Manufacture is based on the autoclave moulding technique. The layers of fibres are pre-impregnated with resin, placed in the cone tooling, and covered by a pressure bag. The assembly is then placed in the autoclave with sufficient heat and gas pressure to cure the resin and consolidate the layers. Once set, the cone is machined around the flange (removing material from sacrificial layers) and the holes are drilled. The cone is then painted, coated with polyurethane, and the rubber parts bonded to it.

Fans

The fan system has two primary functions:

❯ compress the bypass air

❯ feed supercharged air into the core.

In a turbofan, a proportion of the air from the low-pressure compressor passes into the core compression system – the remainder of the air, the bypass flow, is ducted around the core compression system. Both flows eventually pass through separate or integrated propelling nozzles at the rear of the engine to generate thrust.

The civil, high bypass ratio fan has a pressure ratio approaching 2:1. This bypass air expands through the exhaust nozzle and contributes around 75 per cent of the engine

thrust. A low bypass ratio military fan has a pressure ratio typically in the range 3:1 to 4:1. This air passes down the bypass duct, and is then mixed with the core airflow from the turbines, and expanded through the exhaust nozzle. The bypass air is also used for afterburning and to cool the reheat and nozzle system.

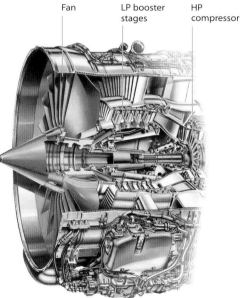

Fan   LP booster stages   HP compressor

Booster stages in the core section on the two-shaft V2500[*1]

The fan system of the Pegasus V/STOL engine in the Harrier is an exception; here, the bypass air is passed directly to the front nozzles of the lift system to generate thrust.

This functionality needs to be achieved at a high level of aerodynamic efficiency, at a low life-cycle cost, weight, and diameter, and at a low level of noise (civil rather than military). The system must also have an adequate stability margin and be able to cope with harsh operating environments.

The system has to pass rigorous certification tests: rain, hail, icing, operability, bird strike requirements, fan-blade-off, any distortion of inlet airflow resulting from aircraft manoeuvres or cross-wind, altitude, and compatibility with intake and thrust reverser. Achievement of noise targets is also of crucial importance.

The fan system must be designed to cope with impact from a range of bird sizes at various positions on the fan face; the size of the bird is a function of intake diameter, so the larger the diameter of the fan intake, the larger the weight of bird that must be accepted. The system has to be able to demonstrate integrity for all types of bird specified in the certification requirements (» 44).

Distortion of inlet airflow is a significant issue for military fans, given the comparatively extreme manoeuvres of military aircraft and the often more complex air intake system. Because of the need for a higher pressure ratio, military fans tend to be multi-stage, and may be configured with VIGVs. The splitter on military fans is usually downstream of the fan bypass vanes, or outlet guide vanes (OGVs).

Civil fans

The modern civil aero engine has a very high bypass ratio turbofan configuration. In this configuration, the intake air undergoes only one stage of compression in the fan before being split between the core (or gas generator system) and the bypass stream. For modern engines, the bypass ratio can be as high as ten. This results in the optimum configuration for passenger and transport aircraft flying at just below the speed of sound. For large engines, a three-shaft configuration is

Wide-chord, swept fan blades on a high bypass turbofan

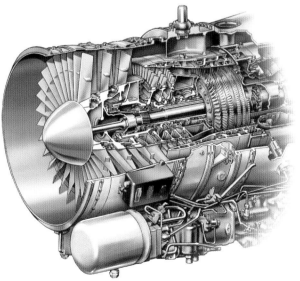

The three-stage military fan on the RB199

preferred, with an intermediate pressure (IP) compressor and high pressure (HP) compressor in the core section.

The major components of the civil fan system are the fan blades, fan disc, containment casing, and the front bearing housing structure containing the bypass vanes and engine section stators.

To reduce the fan diameter, and therefore weight and drag, the inlet hub-tip ratio is minimised, subject to meeting mechanical criteria for the hub design.

The fan blade comprises an aerofoil with a root attachment that secures the blade into the fan disc. The rotor is attached to the fan shaft, which is connected to and driven by the LP turbine. The whole fan rotor assembly is supported by the front bearing housing. The flow leaving the OGVs is axial. The flow leaving the engine section stators may be axial or swirling, depending on the engine configuration.

Fan blade
The hollow, wide-chord fan blade allows higher flow, higher efficiency, and is quieter than its predecessor, the snubbered blade. A snubbered blade consists of a solid aerofoil, which has two appendages, or snubbers, attached at right angles to the aerofoil span at about three quarters of the blade height. These are also known as clappers. When the

blade is assembled, these snubbers form a support, which resists twisting of the aerofoil when subjected to cyclic loading caused by aerodynamic distortion and wakes. They also raise the natural frequencies of the blade and provide a source of damping.

Simply making the blade snubberless results in a design that is too flexible (its natural frequencies are too low) and removes the mechanism for damping any aerofoil vibration. To overcome this, the blade chord is increased, stiffening the blades and allowing a reduction in the number of aerofoils.

One of the principal reasons why civil designs adopted the wide-chord fan is efficiency. Snubbers introduce a significant amount of aerodynamic loss, resulting in a very inefficient design; they also present a blockage to the airflow, requiring the frontal area to be increased.

To avoid excessive fan module weight, the aerofoil is hollow; this not only lightens the individual fan blade but also lightens the whole system (disc, front structure, containment casing).

A hollow blade has a cavity within the aerofoil and is formed from three sheets of titanium: two outer sheets and one inner sheet – a very thin membrane. These blades are produced using diffusion bonding and super-plastic forming processes.

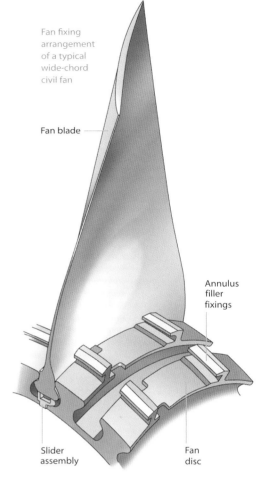

Fan fixing arrangement of a typical wide-chord civil fan

Fan blade

Annulus filler fixings

Slider assembly

Fan disc

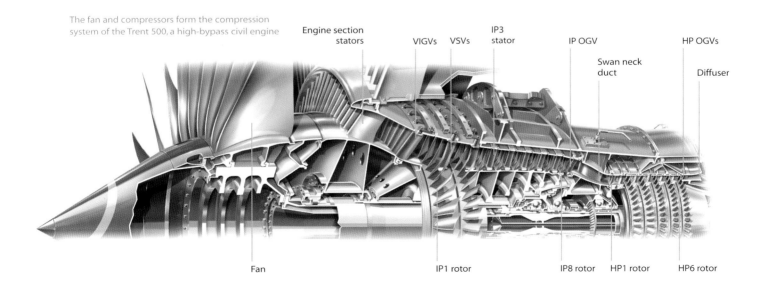

The fan and compressors form the compression system of the Trent 500, a high-bypass civil engine

Engine section stators

VIGVs

VSVs

IP3 stator

IP OGV

Swan neck duct

HP OGVs

Diffuser

Fan

IP1 rotor

IP8 rotor

HP1 rotor

HP6 rotor

Hollow blades behave in a very similar manner to solid blades and there is no detriment in stiffness or bird strike capability. The larger the blade, the greater the benefit from hollow blade technology as more weight can be saved. As blades reduce in size, they can no longer be hollow because the panels would become too thin.

Fan disc

The fan disc is one of the most critical components in the engine and has four main functions:

> react to centrifugal loads from the fan blades – both during normal running and in the event of a fan-blade-off

> provide attachment from the LP shaft to drive the fan and to retain the fan blades

> absorb impact loads

> provide attachment for the nose cone and other peripheral components.

As disc failure is hazardous to the aircraft, this component is classified as a critical part. The disc contains a number of slots into which the fan blades are fitted and there is a front drive arm, which provides attachment to the nose cone assembly. The disc material is usually forged titanium.

The mechanical design of the disc is one of the key design areas, because it is a critical part, and it is an extremely heavy component of the fan system.

The role of the disc is to ensure that the blades continue to travel in a circular path and resist their high centrifugal loads – about one hundred tons, equivalent to ten double-decker buses hanging from each fan blade.

The total disc stress is a combination of inertia stresses of the disc itself and the stresses imposed by centrifugal force on the blades. Two key issues govern the amount of stress the disc is designed to withstand. First, the burst criteria state that, if the assembly overspeeds, the disc will not burst and compromise the integrity of the engine; this provides the minimum cross-sectional area for the disc. The second major issue is the life of the disc; this sets the maximum stress in

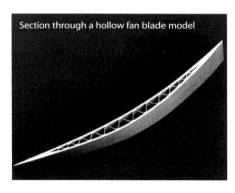
Section through a hollow fan blade model

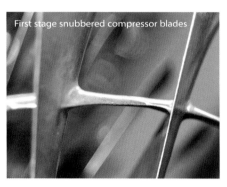
First stage snubbered compressor blades

the disc. If it is unable to meet the life criteria, then various strategies can be employed:

> increase the size of the disc, so that the stress in the disc reduces to acceptable levels. Any extra material added is put at the bore, the most weight efficient location

> decrease or eliminate any stress concentrations in the disc, such as small holes or tight radii

> increase the capability of the material

> if the material properties exceed the life requirement, then the disc size can be reduced until it reaches the minimum size and weight, as specified by the burst overspeed margin.

A Trent fan disc: the force locating a Trent fan blade is powerful enough to throw a car 30 metres into the air

Fan casing

The primary functions of the fan case are to form the outer gas path, and contain a fan blade should it disintegrate during flight. The casing must be capable of absorbing the energy of a complete fan blade, without releasing blade or case fragments, and maintain the integrity of the engine. The energy of a released fan blade is equivalent to a family saloon car at 100kmh (60mph). The casing therefore needs to have high strength and high ductility.

In some engines, the fan case is part of the engine mounting system and thus transmits thrust from the core engine to the aircraft.

It interfaces at its front flange with the nacelle and at its rear flange with the rear case (typically military) or with an OGV ring (typically high bypass civil). The fan case also provides mounts for the gearbox, ground support equipment, and other accessories mounted on the accessories flange. The casing assembly also contains acoustic liners to attenuate noise generated by the fan. The panels are made from a honeycomb structure of composite construction. The fan case inner profile, when fully assembled with the in-fill panels, fan track liner, acoustic panels, and ice impact panel, forms the outer annulus line.

Containment system weight is a function of the fan diameter cubed; so high bypass ratio engines with a large fan blade tip-to-tip diameter have much heavier containment systems.

Military fans

The major components of the military fan module are the rotor drum, casing and other statics, fan blades, and support structures. Military fans can be configured with either a rear support structure (an overhung rotor), or a front and rear support structure (a 'straddle-mounted' rotor). Where a VIGV stage is present, this is incorporated into the front bearing structure.

To minimise the frontal area, and hence weight, drag, and ultimately airframe visibility, the inlet hub-tip ratio is kept as low as possible. The need for low frontal area and low weight means that, in modern military engines, the rotor assembly often has blisks, where the blades and disc are integrated into a single component.

The stators are usually of shrouded construction and are mounted into the casing. The flow leaving the OGV is axial. The casing may also have a casing treatment to improve the fan's stall and surge characteristics.

Fan rotor

The fan rotor configuration has traditionally consisted of two or three discs, each with a set of rotor blades of aerofoil cross-section. The discs and, in more modern engines, blisks, can be bolted or welded together.

The blisk is a challenging component to manufacture. There are two very different methods:

› machining from a single, solid piece of metal

› linear friction welding – this allows the engineer both to optimise the properties of the aerofoil and disc, and also to use hollow aerofoil technology so that the blades and disc can be even lighter.

In the fan system, blades, discs, and blisks are usually made of titanium.

Fan casing and statics

The military fan casing has various functions:

› form the outer gas path, and provide close control of tip clearance for the rotors

› support the statics (stator vanes), and also the front bearing structure

› provide a mount for the VIGV actuation system where present

› provide a containment system for the rotor blades

› mount engine accessories.

The casing needs to have high strength and ductility to achieve these requirements.

The inside surface of the casing incorporates an abradable lining material similar to those used in civil casings. The abradable material is axially aligned with the rotor tips and helps to maintain tight tip clearances, which are critical for performance and stability.

The casing is normally split horizontally in two halves, with the vanes secured into the casing via a dovetail fixing. The vanes are shrouded and are fitted with an inner shroud ring to provide integrity. The inside diameter of the shroud ring incorporates abradable material, which provides a sealing face against the rotor labyrinth seal thereby preventing the leakage of air from stator exit to stator inlet.

The casing and vanes can also be of integral construction: the vanes are built up into rings, like cartwheels, and then assembled to form the casing.

For part-speed operation, VIGVs and VSVs can be used and casing treatments can also be applied. The most common form of casing treatment is circumferential grooves, but slotted style treatments are also used.

Core compressors

The core compressor system has three main functions:

› raise the pressure of the air supplied to the combustor and deliver it at a suitable Mach number with acceptable radial flow properties

› supply bleed air for engine sealing, anti-icing, cooling, and aircraft environmental control

› provide for any power off-take requirements.

Like the fan system, the core compressor system has to demonstrate a high level of aerodynamic efficiency with adequate stability margin for all fan exit conditions, and at a low life-cycle cost and weight. It must also meet similar certification requirements.

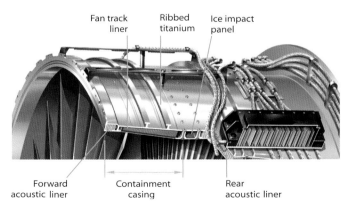

Fan track liner   Ribbed titanium   Ice impact panel

Forward acoustic liner   Containment casing   Rear acoustic liner

Section through a Trent 500 fan casing. This casing mounts the engine accessories, minimises fan noise impact, and must contain a fan blade during a blade-off event.

Configurations

A core compressor system may consist of one or two compressor modules, each driven by its own turbine. Core compressor module pressure ratios are typically in the range from 5 to 16. The core compressor configuration is dependent upon the engine application. The optimum configuration comes from a series of trade-off studies looking at performance, weight, cost, stability, and life.

For large civil engines, the use of two core modules – the three-shaft layout – is usually preferred and provides for a very flexible, robust, and efficient system, allowing each module to run at its optimum rotational speed. It also has the benefit of minimising the number of variable vane stages. Small civil engines and military engines tend to have single core modules – the two-shaft layout.

Axial compressors

A single-spool, axial compressor consists of one rotor and stator assembly containing as many stages as necessary to achieve the desired overall pressure ratio.

The major components of the core module are the rotor drum, casing and other statics, OGV assembly, combustor pre-diffuser, and one or more support structures.

Rotors

The core rotor configuration has typically consisted of 3 to 12 discs, each with a set of rotor blades of aerofoil cross-section. The discs can be bolted or welded together to form an integral drum. Military engines now tend to use rotors of blisk construction to minimise weight. Blisks can also be required where there are space constraints to the disc bore diameter, or where very low hub-tip ratios or hub diameters are required.

Blades, discs, and blisks are made from a range of materials. In modern engines, forward compressor stages are usually made from titanium due to its high strength-to-weight ratio. The rear stages of high overall pressure ratio and military engines are dependent on nickel alloys because of the high operating temperatures.

Blades

The conventional bladed disc solution is typical for civil core compressor designs. Compressor blades are normally attached to the disc using a mechanical feature known as a root fixing. In general, the aim is to design a securing feature that imposes the lightest possible load on the supporting disc thus minimising disc weight. There are two principal fixing methods in use:

> Axial fixing is where a series of slots are machined out of the disc to accept the dovetail or fir-tree shaped rotor blade fixing. Axial fixings are a more complex and costly option; however, they are generally more robust for handling foreign object damage and better facilitate the use of variable vanes. For these reasons, the front stages of a compressor tend to use axial fixings.

> Circumferential fixings are usually the simpler and cheaper option and are common in the rear stages of a compressor. It is relatively easy to machine an annular groove at the head of the disc. Blades are assembled into the disc through a loading slot. The ring is then closed with a locking device.

Blade fixings have the advantage of easy maintenance: damaged blades can be replaced relatively easily. The penalty of using root fixings is that they add parasitic mass, which increases the centrifugal load applied to the disc.

Compressor discs

As with the fan, the mechanical design of the compressor disc is another of the key design tasks. Failure of a disc would seriously compromise the integrity of the engine. In addition, the disc assembly forms a significant fraction of the module weight.

The total disc stress is made up from a combination of the stresses imposed by the blades and spacers, the inertia stresses within the disc itself, and the thermal stresses imposed by bore to rim thermal gradients. These thermal stresses are becoming more significant with increased core temperatures.

Thermal stresses are induced when the rim heats up quicker than the cob (central thickened ring) during acceleration. And also when, from steady state running, the speed is reduced and the cob cools down much more slowly than the rim. Generally speaking, the greater the size of the disc, the less thermally responsive it is and the higher the thermal stress levels.

A Trent HP compressor drum

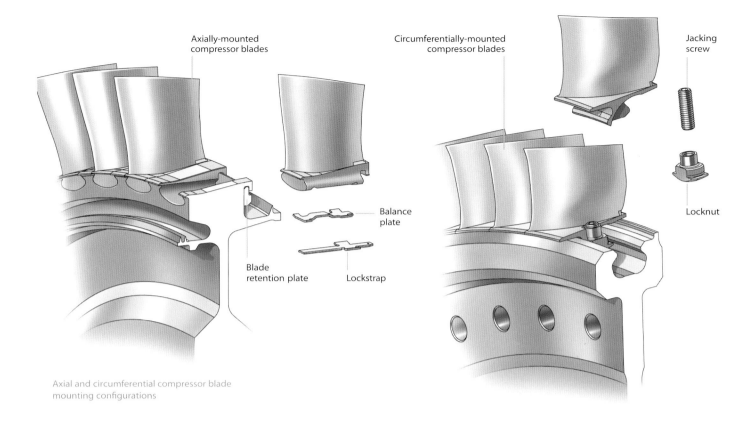

Axially-mounted compressor blades

Circumferentially-mounted compressor blades

Jacking screw

Balance plate

Blade retention plate

Lockstrap

Locknut

Axial and circumferential compressor blade mounting configurations

The mechanical design of the discs in the rear stages of the core compressor is very challenging with three major factors to be considered:

› high rim speed

› high temperatures

› additional loads from the drive arm from the shaft to the turbine.

Further significant reductions in rotor weight may be possible in future through application of 'bling' technology. This would be achieved by replacing current bladed disc and blisk configurations with a high strength reinforced ring, with blades integrated into a single component.

Compressor casing and statics
The core compressor casing has similar functions and requirements to the fan casing. It is designed to tolerate the loads resulting from in-flight manoeuvres. It also needs to have materials that closely match the thermal and centrifugal growth properties to the rotor to achieve acceptable tip clearance under both steady state and transient conditions. Poor control of the matching between the

rotor and casing during transient operation will result in excessive tip clearances and a loss of compressor stability that can result in either stall or surge.

As with the fan, the inside surface of the casing incorporates abradable lining materials, which are thermally sprayed onto the casing. These help to maintain the tight tip clearances that are critical for performance and stability.

There are a number of casing design configurations to be considered: a double-skinned, a single-skinned, and a hybrid casing design. The double-skinned casing configuration is much stiffer than the single-skinned casing and is therefore better able to withstand the loadings resulting from flight manoeuvres. This configuration also provides a simpler solution for bleed off-take. The single-skinned casing is cheaper due to the reduced parts count. There is often little difference in casing weight for these two configurations, as the single-skin casing often has to be thicker in section.

Some compressors use a hybrid design, a combination of these configurations, with

a single skin at the front and double skin at the rear. The casing construction can be split horizontally into two or consist of a number of rings. Typical materials are titanium, steel, and nickel.

The vanes are normally secured into the casing using a dovetail or T-slot fixing, with an anti-rotation feature locking the vanes in position. The vanes are either shrouded or cantilevered.

Shrouded vanes are usually fitted with an inner shroud ring that is secured to the vane shroud by a T-slot fixing. The inside diameter of the shroud ring incorporates an abradable material, which provides a sealing face against the rotor labyrinth seal in order to minimise the leakage of air from stator exit to stator inlet. Shrouded vanes are used in the front stages for their superior resistance to impact.

Cantilevered stators are of much simpler construction and can be used for the mid and rear stages in core compressors. Because of their simplicity and reduced part count, they are cheaper than shrouded vanes. A coating is applied to the compressor drum to protect the vane tip and aid tip clearance.

Typical vane materials are titanium and steel for the forward stages, and nickel for the rear stages. Where variable vanes are present, the casing is also used to mount the actuation system. It may also be necessary to incorporate casing treatment on some stages.

Centrifugal compressors

Centrifugal compressors generally comprise four major subcomponents: inlet duct, impeller, radial diffuser, and exit system. To achieve higher pressure ratios, centrifugal compressors may be 'staged' with multiple compressors in series.

The inlet duct may be either radial or axial in shape, and may incorporate pre-swirl vanes to provide an initial swirl to the air entering the compressor impeller.

Impellers

The impeller consists of a forged or cast disc with integral, radially disposed vanes on one or both sides forming convergent passages in conjunction with the compressor casing. Impellers can either be of single- or double-sided configuration, and may incorporate partial vanes, or splitters. These splitters are located part way down the vane passage, and extend to the impeller exit plane. To ease the air from axial flow in the entry duct onto the rotating impeller, the vanes in the centre of the impeller are curved in the direction of rotation. The curved sections may be integral with the radial vanes or formed separately for easier and more accurate manufacture.

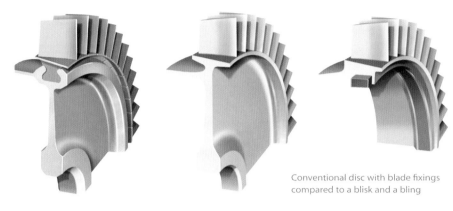

Conventional disc with blade fixings compared to a blisk and a bling

Radial diffusers

The diffuser assembly may be an integral part of the compressor casing or a separately attached assembly. It may consist of a number of vanes formed tangentially to the impeller, a number of intersecting conical drillings (pipes), or no passages at all – just divergent walls (vaneless). The passages are divergent, to convert the kinetic energy into pressure energy. The inner leading edges of the passages are in line with the direction of the resultant airflow from the impeller.

The clearance between the impeller and the diffuser is an important factor. Too small a clearance will set up aerodynamic buffeting impulses that could be transferred to the impeller and create an unsteady airflow leading to vibration, which may be mechanically destructive. Too high clearance will decrease compressor efficiency.

Exit systems

Centrifugal compressor exit system geometries are usually dictated by the engine general arrangement. They may have single or multiple exit collecting scrolls, an annular bend from radial to axial followed by an axial de-swirl cascade, or may dump to a plenum. The function of the exit system is to minimise the exit pressure loss, while performing further diffusion, and to align the air direction required for the following engine components.

Industrial and marine compressors

Aero-derivative engines remove the fan entirely. They either make the IP compressor the engine inlet, or replace the fan with a new LP compressor. For example, when a Trent core is used on the Industrial Trent, the fan is removed and replaced with a two-stage LP compressor linked to the LP turbine.

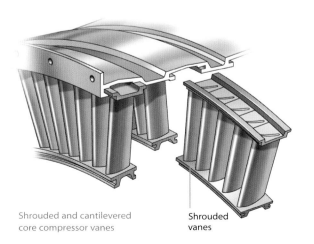

Shrouded and cantilevered core compressor vanes

Shrouded vanes

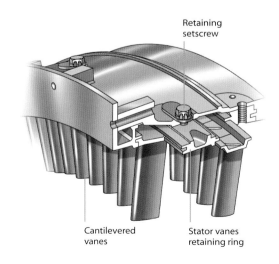

Retaining setscrew

Cantilevered vanes

Stator vanes retaining ring

The MT30 marine engine does not have a two-stage LP compressor, and air is induced straight into the aero-derivative IP compressor.

Potential modifications to the IP compressor are:

> Redesign of the first stage to reflect the absence of the 'hub low' inlet pressure profile caused by the fan. This increases the pressure ratio towards the tip, and hence the compressor inlet flow; Historically, around six per cent flow increase has been achieved.

> Coatings or material changes are often necessary to prevent corrosion in an offshore industrial or marine environment.

Compressor rigs

Compressor rigs are used to acquire new capability, and to support engine development programmes where required. These can be rigs that are fully representative of the actual engine hardware, and are operated at high-speed or low-speed representations of a stage.

Low-speed rigs have the advantages of much lower cost and greater physical size, but operate at flow conditions (specifically Mach number) that are significantly below the actual operating environment. These types of rigs are not suitable for replicating stages that have flow above subsonic air velocities.

Much effort has gone into the development of new methods, in particular 3D CFD capability. Confidence in these methods has grown significantly in recent years, and consequently the use of engine development rig test vehicles has diminished or in some cases been completely eliminated. This has had a significant impact on the development process through reductions in timescales and costs for new engine programmes.

The future

The challenges for the compression system will continue with further engine cycle demands for increased bypass ratio, overall pressure ratio, efficiency, and delivery temperature. At the same time, the requirement for low life-cycle cost, weight, and noise will become ever more challenging, while development costs and timescales must be further reduced.

This drives the need for research into improved aerodynamics and mechanics, into materials that weigh less but can tolerate higher operating temperatures, and into the manufacturing technology developments needed to turn these new materials into actual engine components.

There will be a strong focus on the ability to accurately model all the significant attributes (life-cycle cost, weight, performance, and noise) of the compression system, with the ultimate aim of being able to design the optimum compression system for all engine applications.

There will be a trend towards much larger and more sophisticated models, with much more of the overall system and surrounding environment within the calculation.

The current wide-chord fan blade family. Fan diameter, for a given thrust, will continue to increase

As the air leaves the core compressor, it is travelling at around 150 metres per second. Aviation fuel cannot burn in this environment.

combustors

THE JET ENGINE IS A HEAT ENGINE, AND THE COMBUSTOR IS WHERE THAT HEAT IS CREATED BY CONVERTING THE CHEMICAL ENERGY OF THE FUEL INTO THERMAL ENERGY. HISTORICALLY, THE COMBUSTOR HAS ALWAYS BEEN ONE OF THE MOST DIFFICULT AREAS OF THE ENGINE TO GET RIGHT.

combustors

The combustion chamber has the difficult task of burning large quantities of fuel with extensive volumes of air from the compressor. Heat must be released in such a way that the combustion gases are expanded in a smooth stream of uniformly heated gas – while also meeting the following requirements:

A close up view of a combustor tile (top) and the same tile in situ (bottom)

》 high combustion efficiency to ensure maximum heat release

》 wide range of stability so that the flame stays alight even when the engine ingests large quantities of rain or hail, and during rapid decelerations

》 reliable ignition on cold days

》 ability to restart the engine and pull away at high altitude

》 low pressure loss in order to maximise overall engine performance, but sufficient pressure loss to drive cooling air through the turbine

》 a temperature profile at the combustor exit that matches the life requirements of the turbine

》 low emissions, especially for some industrial engines

》 high durability for reliability, long life, and to minimise maintenance

》 low cost

A modern airspray nozzle

❭ low weight, particularly for aero engines in order to achieve

  ❭ lower fuel consumption

  ❭ greater load-carrying capacity

  ❭ a high thrust to weight ratio in military aircraft

❭ ability to burn a wide range of fuels:

  ❭ aero engines burn kerosene

  ❭ marine engines burn diesel

  ❭ industrial engines may burn both these plus natural gas of varying composition.

There is a very fine balance to ensure that each one of these design requirements are met. The performance of the combustor often hinges on subtle changes to the admission of air, the fuel injector, and the cooling features. Changes made to improve one aspect invariably have an impact, often adverse, elsewhere. In addition to the unique aerothermal challenges, the high-temperature, high-pressure, and high-vibration environment provides particularly difficult mechanical integrity challenges.

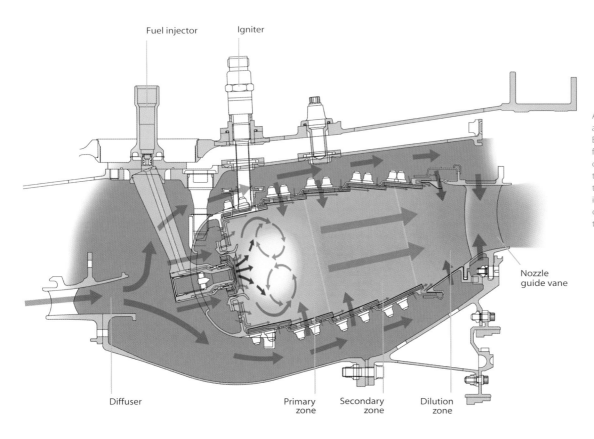

Fuel injector

Igniter

Air and fuel flow through
an annular combustor.
Blue shows the combustion
feed air from the HP
compressor, and white
through yellow to red,
the hot combustion gases
in the primary zone being
cooled before entering
the turbine system.

Nozzle
guide vane

Diffuser

Primary
zone

Secondary
zone

Dilution
zone

The combustion process

For a large civil aero engine, air may leave
the compressor at a velocity of approximately
150 metres per second. This is far too high
a speed for combustion to occur so the air
passes through a pre-diffuser at the front of
the combustion module, reducing the axial
velocity to about 110m/s. This is still too
high a velocity for a flame to stabilise as
the flame front of burning kerosene has
a velocity of only 10m/s. A dump diffuser
may considerably reduce velocity around the
outside of the combustor, but, as the air
enters the combustor through the mixing
ports, its velocity is still approximately 100m/s.
Stable combustion can only be maintained by
creating lower velocity recirculation regions
immediately downstream of the fuel spray
nozzle. The section of the combustor in
which this recirculation occurs is known
as the primary zone.

The conical fuel spray from a fuel spray nozzle
intersects the recirculation vortex in the
primary zone. This action, together with the
general turbulence in the primary zone,

promotes the break-up of fuel and mixing
with the air, both of which are necessary to
ensure high combustion efficiency and low
emissions. An electric spark from an igniter
plug positioned in the primary zone initiates
the flame that must then be self-sustaining.

The temperature of the gases released
by combustion is approximately 2,100°C.
This is too hot for entry to the nozzle guide
vanes (NGVs) and first rotor blades of the
turbine system, so, in order to reduce the
gas temperature, more air is introduced
into the secondary zone of the combustor
downstream of the primary zone. This air,
which enters the secondary zone through
intermediate ports, also plays a key role in
controlling emissions. Finally, in the dilution
zone towards the rear of the combustor, more
air is introduced to control the temperature
profile of the gases at the combustor exit.
Combustion should be completed before
the dilution air enters the combustor, or the
incoming air will cool the flame. This would
mean that combustion would continue in the
downstream components causing overheating.

Scaling, loading,
and combustion efficiency

Engine components are frequently scaled
to match them to differing operating cycles
but the combustor is the least amenable to
scaling. Combustor loading is a parameter
against which operational parameters such
as efficiency, relight, and pull-away may be
predicted and can therefore be employed
to scale the volume of a combustor.
The loading parameter is proportional
to mass flow, but inversely proportional
to combustor inlet pressure, velocity,
and the total inlet temperature.

Combustion efficiency is effectively 100 per
cent at take-off conditions; it reduces at
lower temperatures and pressures, with
increasing loading parameter. To avoid the
production of 'white' smoke, efficiency must
be retained above about 96 per cent and at
no point in the operating cycle is less than
90 per cent acceptable.

Following a flame-out at altitude, the
combustor must be able to relight and pull

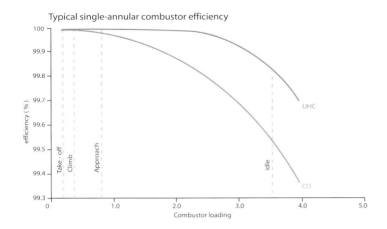

Typical single-annular combustor efficiency

Combustion efficiency for unburnt hydrocarbons (UHC) and carbon monoxide (CO) against the combustor loading parameter

away in a sufficiently short time to enable the engine to regain sufficient spool speed to continue flying. This is dependent on the rate at which the flame can propagate and generate heat, and therefore is directly related to combustion efficiency at low power conditions. As pressure decreases, the combustor volume to achieve a given loading must be increased; therefore, altitude ignition and pull-away are the key performance parameters to sizing the combustor volume.

A multiple combustion chamber system with individual flame tubes and air casings

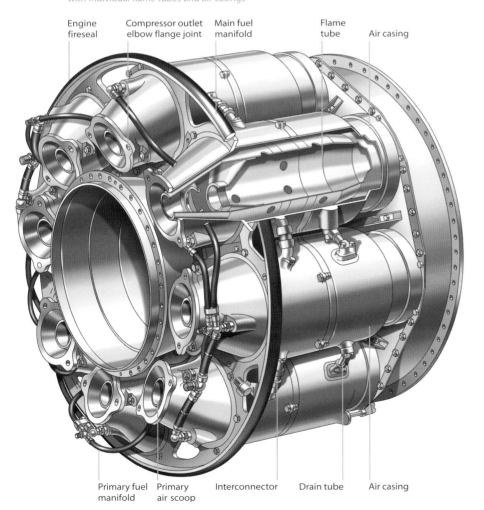

Engine fireseal

Compressor outlet elbow flange joint

Main fuel manifold

Flame tube

Air casing

Primary fuel manifold

Primary air scoop

Interconnector

Drain tube

Air casing

Combustion module architecture

There are three main types of combustion chamber used for gas turbine engines: the multiple chamber, the annular chamber, and the tubo-annular chamber.

Multiple combustion chamber system

The multiple combustion chamber system is made up of a series of individual chambers positioned around the engine. Each chamber has an inner flame tube with its own air casing. Ducts direct air from the compressor into each chamber. The air passes through the flame tube snout and also between the tube and the outer air casing. The separate flame tubes are normally all interconnected; this allows combustion to propagate around the flame tubes during engine starting, and also means that the tubes operate at the same pressure.

This layout is a development of the early type of Whittle combustor. It is no longer favoured for aero applications, but is used in some industrial applications. On early aero engines with this layout, the chambers were aligned parallel to the engine centreline, but industrial engines with multiple combustion chambers may position the chambers perpendicular to the engine centreline. This architecture also reduces the time taken to maintain the combustors, and can accommodate a larger combustor, which may be needed to control emissions. Testing during development is also simpler with this layout as much of it can be done with just one chamber. (» 47)

Tubo-annular combustion chamber

The tubo-annular combustion chamber evolved from the multiple chamber system and paved the way for the annular type. A number of flame tubes are fitted inside a common air casing. The airflow is similar to that in the multiple combustion chambers, but not all the air enters the front of the tube; a significant amount enters through the side wall of each flame tube in a manner similar to the annular system. The tubo-annular arrangement combines the ease of overhaul and testing of the multiple system with some of the compactness of the annular system.

Annular combustion system

This type of combustion chamber consists of a single flame tube, annular in form, which is contained in an inner and outer casing. The airflow is again similar to that already described, the chamber being open at the front to the compressor and at the rear to the turbine nozzles. This style of combustor is predominant in modern gas turbines.

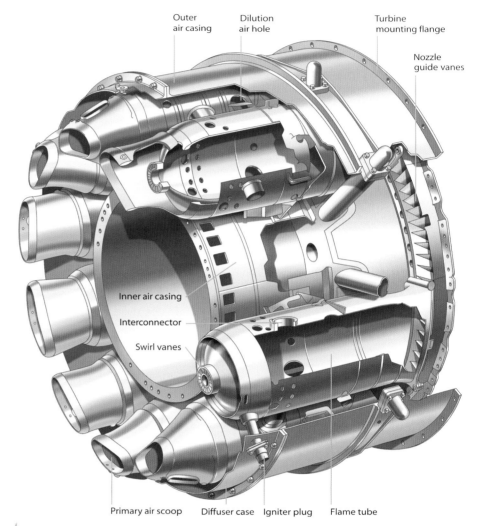

The tubo-annular combustion system has individual flame tubes, but a common air casing

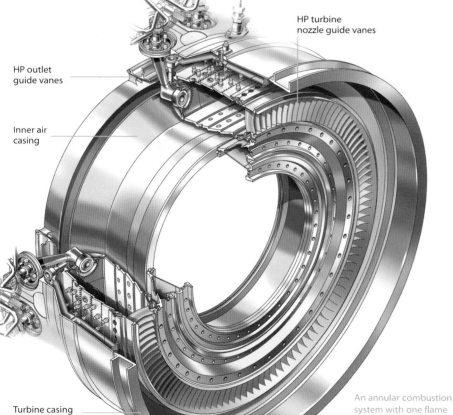

An annular combustion system with one flame tube and one air casing

The main advantage of the annular chamber is that, for the same power output, the length of the chamber need only be 75 per cent of a tubo-annular system of the same diameter. This results in a shorter, stiffer engine and a considerable saving in weight and production cost. An annular combustor will also have a smaller frontal area than a tubo-annular combustor of the same volume. Another advantage is the elimination of combustion propagation problems from chamber to chamber.

An annular combustor has a smaller wall area than a comparable tubo-annular combustion system and requires about 15 per cent less cooling air to prevent burning of the flame tube. This air can instead be used in the combustion process, helping increase combustion efficiency and control emissions.

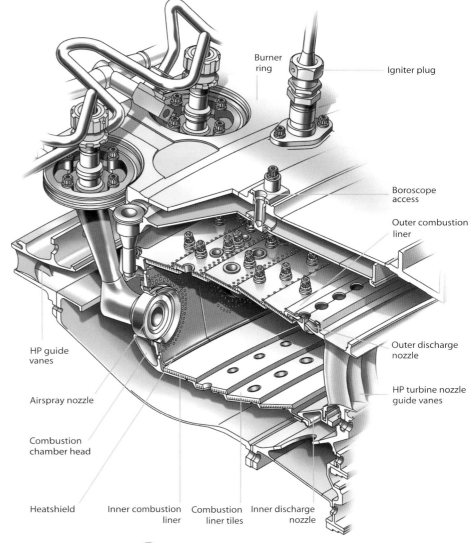

Section through an annular combustion chamber showing an airspray nozzle

Disadvantages of the annular system are that it is structurally weaker, more complex to manufacture, and it is more difficult to control the exit temperature of the gases. Development testing is also more complex. This testing is preferably carried out on the complete system, but time and cost restrictions sometimes necessitate testing with sector combustor rigs using four spray nozzles instead of the twenty, for example, in a complete combustor. This significantly reduces the energy required to simulate engine operating conditions.

Burner ring

Igniter plug

Boroscope access

Outer combustion liner

Outer discharge nozzle

HP turbine nozzle guide vanes

HP guide vanes

Airspray nozzle

Combustion chamber head

Heatshield

Inner combustion liner

Combustion liner tiles

Inner discharge nozzle

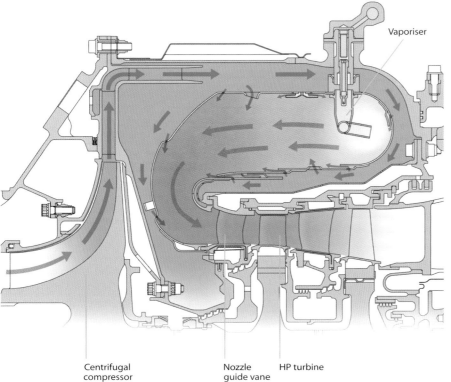

Vaporiser

Airflow through the reverse-flow annular combustion chamber of the RTM322

Annular combustion chambers may have either a straight-through or reverse-flow design. Reverse-flow combustors are particularly compatible with centrifugal compressors and allow the distance between the compressor exit and turbine nozzle entry to be about a third of that required for an equivalent axial combustor. This design approach can produce a very compact engine – critical for helicopter engines. The RTM322 and Gem engines, for example, both have reverse-flow combustors.

Centrifugal compressor

Nozzle guide vane

HP turbine

119

Fuel injectors

The fuel has to be delivered to the combustion chamber where it is thoroughly mixed with air before combustion. For liquid fuels, there are two distinct methods of doing this: vaporisers and fuel spray nozzles, the latter comprising the two main types of pressure-jets and airspray injectors.

Vaporisers

Vaporisers are comparatively simple, cheap, and lightweight structures that serve to mix the fuel and air. Fuel is injected through a fuel-feed tube or sprayer into an L- or T-shaped tube that turns the fuel/air mixture through 180 degrees. The corners of the vaporiser are typically sharp and are intended to create vortices and promote mixing. These may be supplemented by weirs inside the vaporiser, which also encourage turbulence and mixing. Although the fuel/air mixture is heated inside the vaporiser, most of the mixture leaves the vaporiser and impinges on the combustor baseplate as a series of droplets that

receive heat and are vaporised by the high temperatures in the primary zone of the combustor. Some combustor designs require the addition of specialised air feed features such as 'blown rings' to blow fuel away from the walls to improve efficiency. Engines with vaporisers additionally require primers, which are pressure-jet fuel injectors, to improve ignition characteristics by delivering atomised fuel near the igniters.

The vaporiser is fuel-cooled and has a tendency to overheat when the engine decelerates because the combustion gases in the primary zone are still radiating and conducting heat, but there is little fuel to cool the vaporiser. Because it is fuel-cooled, the vaporiser is also susceptible to overheating caused by blockage of the fuel feed tube.

Vaporisers have been predominant in applications requiring simple, cheap, and lightweight fuel injectors, particularly military aero engines like the Pegasus and RB199,

and the RTM322 and Gem helicopter engines. They were also used in the Olympus 593 that powered Concorde. They have not been favoured on large civil aero engines because of durability and emissions requirements.

While vaporisers are able to offer high efficiencies and can give low smoke at reasonably high pressures, they are unable to produce satisfactorily low smoke at the very high temperatures and pressures seen in the latest generation of civil and military high-thrust aero engines.

Fuel spray nozzles

The fuel spray nozzles atomise the fuel to ensure its rapid evaporation and burning when mixed with air. This combustion is a difficult process for two reasons: the velocity of the air stream from the compressor creates a hostile environment for the flame, while the short length of the combustion system means there is little time for burning to occur.

Pressure-jet injectors

One technique of atomising the fuel is to pass it through a swirl chamber where tangential holes or slots impart swirl to the fuel. The fuel is then passed through the discharge orifice, where the fuel is atomised to form a cone-shaped spray. This is called pressure-jet atomisation. The rate of swirl and pressure of the fuel at the fuel spray nozzle are important factors in good atomisation. The shape of the spray is an indication of the degree of atomisation: at low fuel pressures, a continuous film of fuel is formed, known as a 'bubble'; at intermediate fuel pressures, the film breaks up at the edges to form a 'tulip'; at high fuel pressures, the tulip shortens towards the orifice and forms a finely atomised spray.

The simplex spray nozzle is a pressure-jet atomiser with a single fuel manifold. Used on early jet engines, it consists of a chamber that induces a swirl into the fuel and a fixed-area atomising orifice. This nozzle gave good atomisation at the higher fuel flows (at high fuel pressures) but was very unsatisfactory at the low pressures required at low engine speeds and especially at high altitude. The simplex is, by the nature of its design, a 'square law' spray nozzle; that is, the flow through the nozzle is proportional to the square of the pressure drop across it. This meant that if the

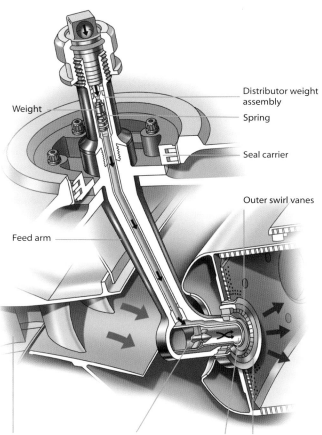

Weight

Distributor weight assembly

Spring

Seal carrier

Outer swirl vanes

Feed arm

HP compressor   Inner swirl vanes   Swirl chamber   Nozzle head

A section through an airspray nozzle

minimum pressure for effective atomisation were 200kPa, the pressure needed to give maximum flow would be about 40,000kPa. The fuel pumps available at that time were unable to cope with such high pressures.

The duplex and duple fuel spray nozzles require a primary and a main fuel manifold and have two independent orifices, one much smaller than the other. The smaller orifice handles the lower flows; the larger deals with the higher flows as the pressure increases. A pressurising valve may be employed with this type of spray nozzle to apportion fuel to the two manifolds. As the fuel flow and pressure increase, the pressurising valve moves to admit fuel progressively into the main manifold and the main orifices. This combined flow down both manifolds allows the duplex and duple fuel spray nozzles to give effective atomisation over a wider flow range than the simplex spray nozzle for the same fuel pressure. The duple has two fuel chambers and two orifices, whereas the duplex has one fuel chamber and two orifices.

Airspray nozzles
The airspray nozzle uses compressor discharge air to create a finely atomised fuel spray.

By aerating the spray, the local fuel-rich concentrations produced by other types of spray nozzle are avoided, giving a reduction in both carbon deposition and exhaust smoke. The airspray fuel spray nozzle will typically have two or three air swirler circuits: an inner, an outer, and a dome. An annular fuel passage between the inner and outer air circuits feeds air onto a prefilming lip. This forms a sheet of fuel that breaks down into ligaments. These ligaments are then broken up into droplets within the shear layers of the surrounding highly swirling air.

The fuel spray nozzle designer not only has to consider optimising the atomisation of fuel, but also where the fuel droplets are directed. These characteristics can be fine-tuned by altering the quantities of air that pass through each air circuit and the amount of swirl that is imparted. An additional advantage is that the low fuel pressure required for atomisation permits the use of the comparatively light gear-type pump.

Fuel distribution
For larger diameter combustion chambers, a flow distributor valve is often required to compensate for the gravity head across the manifold at low fuel pressures to make sure that all the spray nozzles pass an equal quantity of fuel especially at ignition conditions. This ensures that all sectors of the combustor operate in the same way, giving repeatability in the temperature distribution seen by the high pressure (HP) turbine. Small diameter combustion chambers, such as those used on military engines, do not have flow distributor valves, but may nevertheless have to cope with an irregular distribution of fuel pressure caused by high-g manoeuvres (» 175, 179).

Industrial and marine fuel injectors
Industrial engines have an additional complication in that they may be required to run on both liquid and gaseous fuels. This is approached in different ways, depending upon how quickly the change-over is required: 'dual fuel' combustion systems have a single set of fuel injectors and can switch between fuels while running; 'double fuel' combustion systems require the swapping of fuel injectors when fuels are changed. Dual fuel nozzles are evolved from aero liquid-fuel spray nozzles; gas-only fuel injectors operate at lower pressures, and some may use a series of plane orifices to impart swirl to the fuel flow.

Igniters

There are two basic types of igniter plug; the constricted or constrained air gap type and the shunted surface discharge type. The air gap type is similar in operation to the conventional reciprocating engine spark plug, but has a larger air gap between the electrode and igniter body for the spark to cross. A potential difference of approximately 25,000 volts is required to ionise the gap before a spark will occur. This high voltage requires very good insulation throughout the circuit. The surface discharge igniter plug has the end of the insulator formed by a semi-conducting pellet, which permits an electrical leakage from the central high-tension electrode to the body. This ionises the surface of the pellet to provide a low resistance path for the energy stored in the capacitor. The discharge takes the form of a high intensity flashover from the electrode to the body and only requires a potential difference of approximately 2,000 volts for operation.

The normal spark rate of a typical ignition system is between 60 and 100 sparks per minute. Periodic replacement of the igniter plug is necessary due to the progressive erosion of the igniter electrodes caused by each discharge.

The igniter tip has a range of immersions into the combustor flame tube of plus or minus one millimetre depending on flame tube design and wall cooling technology. During operation, the spark penetrates a further 20mm. The fuel mixture is ignited in the relatively stable boundary layer; the flame then propagates throughout the combustion system. A modern annular combustion system usually has two igniters on opposite sides of the annulus.

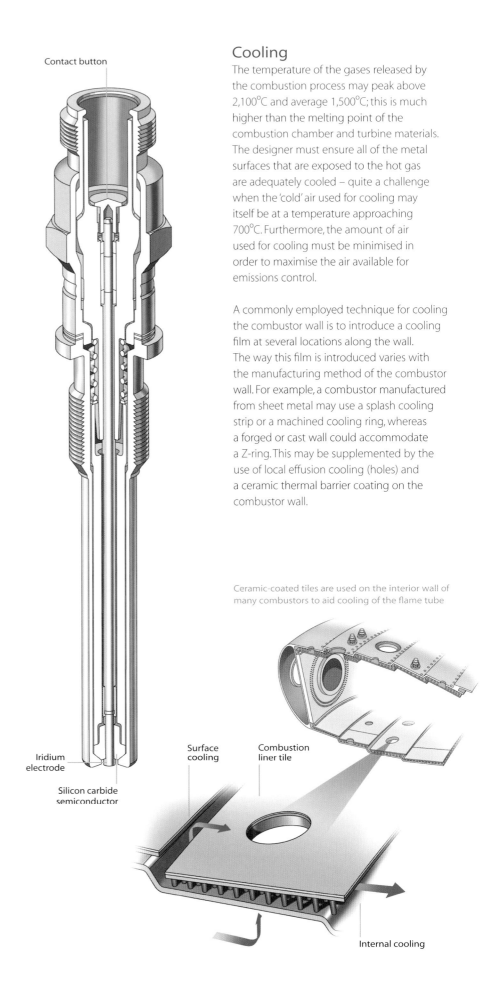

Contact button

Iridium electrode

Silicon carbide semiconductor

Surface cooling

Combustion liner tile

Internal cooling

Ceramic-coated tiles are used on the interior wall of many combustors to aid cooling of the flame tube

Cooling

The temperature of the gases released by the combustion process may peak above 2,100°C and average 1,500°C; this is much higher than the melting point of the combustion chamber and turbine materials. The designer must ensure all of the metal surfaces that are exposed to the hot gas are adequately cooled – quite a challenge when the 'cold' air used for cooling may itself be at a temperature approaching 700°C. Furthermore, the amount of air used for cooling must be minimised in order to maximise the air available for emissions control.

A commonly employed technique for cooling the combustor wall is to introduce a cooling film at several locations along the wall. The way this film is introduced varies with the manufacturing method of the combustor wall. For example, a combustor manufactured from sheet metal may use a splash cooling strip or a machined cooling ring, whereas a forged or cast wall could accommodate a Z-ring. This may be supplemented by the use of local effusion cooling (holes) and a ceramic thermal barrier coating on the combustor wall.

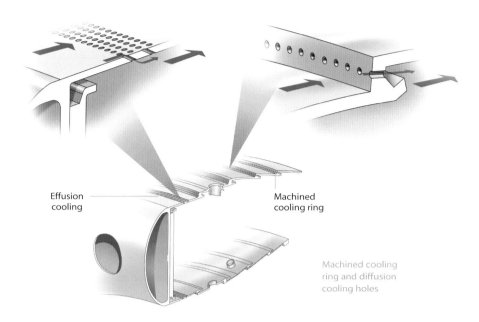

Effusion
cooling

Machined
cooling ring

Machined cooling
ring and diffusion
cooling holes

Cooling air in

Laminated flame
tube wall

Film of cooling air out

Transpiration cooling

Transpiration cooling
uses laminated
materials with a
network of internal
air passages

Many combustors employ ceramic-coated tiles to line the combustor wall. The individual tiles are attached to a cold 'skin', and cooling air passes through holes in the combustor wall and impinges on the tile. The air then moves through a series of pedestals designed to improve the convective heat transfer coefficient, before exiting the front and rear of the tile to form an insulating film. The tiles are designed to be removable for maintenance.

An alternative cooling technique, called transpiration, is to use laminated materials that allow cooling air to enter a network of passages within the flame tube wall before exiting to form an insulating film of air.

The thermal management of fuel-wetted surfaces within the fuel injector is a particular concern. If fuel is exposed to excessive temperatures within the fuel injector, it will decompose to form lacquers and carbon deposits that may block fuel passages or cause distortion. For this reason, the fuel injectors feature complex heat shielding and are carefully designed to prevent regions of stagnant fuel from occurring.

This issue can be more of a problem for industrial and marine applications, where the liquid diesel fuels have lower thermal stability. Subtle combustor cooling changes may also be necessary for industrial and marine applications due to the increased radiation caused by diesel fuel properties.

Predictive modelling

The modelling of metal temperatures is necessary to determine the displacement, thermal stresses, and life of a component. This modelling is done using finite element analysis. In order to calculate metal temperatures, it is necessary to input material property data, engine performance data, air system data, and heat transfer coefficients. These heat transfer coefficients may be validated by computational fluid dynamics (CFD) analysis and/or rig or engine thermocouple measurements. CFD can also allow the designer to model, first, the flow of air in, through, and out of the combustor, second, the complicated air/fuel mixing, and third, the chemistry behind the combustion process.

Testing

In order to develop a combustor that meets all the operational parameters throughout the engine operating range, it is important to test at the relevant conditions. Although the final confirmation of performance will always be in the engine with appropriate turbomachinery, the development programme uses combustion rigs that enable parametric control of the inlet parameters for full evaluation of the combustor performance. This requires a series of test facilities to cover the low to high power parameters:

❭ Combustor airflow distribution and cold pressure loss may be measured on the full combustor hardware at isothermal conditions or more detailed diagnostics can be applied on a perspex model, which simulates all the airflow. This represents validation of the initial semi-empirical design rules employed from diffuser exit to NGV inlet.

❭ The combustor exit temperature traverse pattern that will be presented to the HP NGV and turbine is measured in the combustor exit plane. Traversing thermocouples can measure radial and circumferential temperature distribution, but, for higher temperatures, gas sampling probes, which calculate the gas temperature from the measured gas composition, may be used. This is normally done in a fully annular combustor or as representative a set of tubo-annular combustors as possible.

❭ Emissions are measured across the operating range: CO and UHC are highest at low powers; NO_x and smoke at high powers. All must be compliant with legislation to achieve engine certification. It is preferable to do all measurements in full combustor geometries, but costs of providing engine-level mass flows of air up to 700°C and 5MPa (725psi) may be prohibitive, so multi-sector combustor rigs may be employed, using the central sectors only for analysis to exclude side-wall effects.

❭ Ignition, light-round, pull-away, and weak extinction are measured in fully annular rigs at either sea-level-static or sub-atmospheric rigs to simulate the relevant combustor inlet conditions.

Mechanical integrity

In designing a combustion system, considerable effort is put into ensuring the mechanical integrity of all the components in the module. Predicting component life is an essential part of reliability and service warranties.

Materials

The containing walls and internal parts of the combustion chamber must be capable of resisting the very high gas temperature in the primary zone. In practice, this is achieved by using the best heat-resisting materials available, the use of high heat-resistant coatings, and by cooling the inner wall of the flame tube.

Nickel alloys predominate throughout the combustion module where medium-to-high-strength wrought alloys are used for structural components. Cast nickel alloys are also employed, especially where precision forms are required.

Casings

There are several key elements involved in ensuring the mechanical integrity of the casings: pressure containment, life, fan-blade-off, and shock loads.

Pressure containment

The casings must neither buckle nor rupture under the most extreme pressure loadings seen by the engine. The ability of the casings to withstand the pressure loads is assessed through pressure-vessel tests.

Life

The casings may be required to last the lifetime of the engine, which can vary from 13,000 hours for a naval marine engine, 25,000 flights for a large civil aero engine or 100,000 hours for an industrial engine. The component

The materials exposed to the hot gases of the combustor must be adequately cooled when the air used for cooling can be 700°C

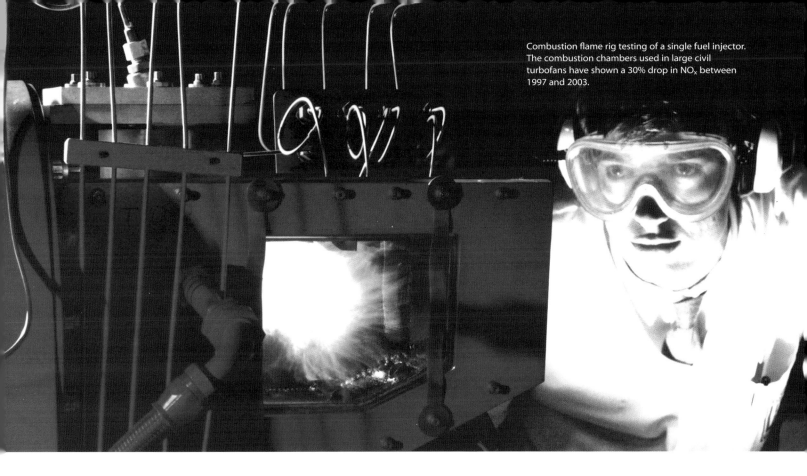

Combustion flame rig testing of a single fuel injector. The combustion chambers used in large civil turbofans have shown a 30% drop in NO_x between 1997 and 2003.

life can be assessed by using finite element analysis to look at the stresses within the casing, with particular attention being paid to local stress concentration features. The boundary conditions within these models must take into account the pressure loads, carcass loads, and thermal effects.

Fan-blade-off (aero engine specific)

The casings must be able to cope if a fan blade is lost during engine running. When this occurs, the shafts decelerate extremely quickly putting large torques and bending moments through the casing. This is replicated by fan-blade-off testing during the engine development programme. Under these conditions, the casing must not buckle and the flange integrity must be maintained. The loads, however, are likely to cause plastic deformation of the metal around the flange bolt holes. Integrity must be maintained despite the large vibration caused by the out-of-balance of the engine.

Shock loads (naval specific)

The casings (and combustor mountings and flanges) must be able to withstand the high bending loads caused by the explosion of a mine or depth charge. As with the loss of a fan blade, the flanges might be expected to deform but must not buckle. Unlike the

fan-blade-off case, the casings do not need to contend with high cycle fatigue caused by engine out-of-balance. However, the engine is expected to operate for a limited period after the shock loading with little performance deterioration.

Combustor

It is necessary to predict when cracks or holes in a combustor will be initiated. Finite element analysis can be used to assess the stress and strain ranges that will be caused by thermal effects and vibration, and these ranges can, in turn, be used to predict crack propagation rates. Unlike casings, however, where cracks cannot be tolerated, cracks may be permitted in combustors, depending on where they occur. Prediction of crack propagation rates is therefore very important. It is also necessary to be able to predict how quickly thermal oxidation will lead to crack initiation.

The input for the finite element models will come from rig and engine thermocouple, strain gauge, and thermal paint data. Thermal paints change colour to indicate the highest temperatures seen by a component and so give a good overall coverage, but thermocouples are necessary to provide 'live' temperatures during running in a rig or engine. However, the assessment of

component lives does not rely solely on these models: it is also determined by cyclic and endurance engine testing.

It is also necessary to consider extraordinary cases. For example, in aerospace applications, the combustor must withstand any loads caused by bird ingestion. This plays a role in determining the number of combustor mounting points. The combustor must also be able to cope with a flame-out – a situation where stable combustion can no longer be maintained and the flame is extinguished. When this occurs, there is still high pressure on the outside of the combustor wall where air is being delivered from the compressor and the pressure inside the combustor rapidly collapses. This puts a buckling load on the combustor outer wall. Engine surge also presents a similar load case.

The challenge of ensuring that the combustor meets its life requirements is made more difficult in the case of marine engines and industrial engines running offshore because of the corrosion caused by salt ingestion and by the high sulphur content of diesel fuels, compared to kerosene. Therefore, in order to combat this corrosion, these engines need slightly different coatings from those used in aero engines.

Flame temperature against air/ fuel ratio

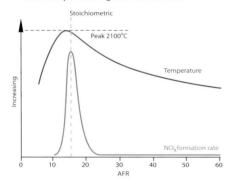

Influence of temperature on CO and NO$_x$ emissions

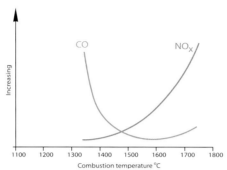

Flame temperature and nitrogen oxide (NO$_2$) formation against AFR. Peak NO$_2$ formation is at the stoichiometric AFR ≈ 15:1, while peak flame temperature occurs at just below stoichiometric.

NO$_x$ and CO emission against flame temperature. The aim of the combustor is to operate in the dip between the two graphs for as much of the engine cycle as possible.

The challenges of designing a clean combustion system

Aerospace considerations

Although the combustor must primarily be designed to ensure stable combustion, the need to control emissions has been the major influence in recent years for the design of the combustor. Bodies such as the International Civil Aviation Organization (ICAO) produce legislation covering the emission of oxides of nitrogen (NO$_x$), carbon monoxide (CO), unburnt hydrocarbons (UHC) and smoke. Further emissions requirements may be placed by the airframer, and also by the end customer. While emissions legislation is becoming increasingly stringent, engine design trends, which have led to richer air/fuel ratios and higher temperatures and pressures inside the combustor, make the control of NO$_x$ and smoke more difficult.

Kerosene is burned efficiently and has the greatest heat release at a mixture strength of about 15 parts of air to 1 part of fuel – an air/fuel ratio or AFR of 15. This is, for kerosene, the stoichiometric ratio in that it enables all of the fuel to burn using all the oxygen in the air. The AFR is the basic parameter that determines the combustor's temperature rise.

However, the mixing of fuel and air within the combustor is not uniform. There are regions near the fuel injector, for example, where the AFR will be richer, but also areas where it will be considerably weaker, with AFRs reaching 130:1 at times. The production of emissions is controlled by the selection of AFR in different

zones of the combustor, but a balance between conflicting requirements must be achieved. For example, the high temperature conditions that help consume smoke are the same as those that generate high NO$_x$ due to the dissociation of atmospheric nitrogen.

The approach taken to optimise emissions for many engines is to burn initially at very rich AFRs to minimise smoke and NO$_x$ production; air is then introduced rapidly through the dilution ports to weaken the AFR to a point where NO$_x$ production ceases but smoke is still consumed.

In addition to satisfying emissions requirements, the temperature profile at the combustor exit, both in a radial and circumferential direction, must be carefully controlled so that it meets the requirements of the turbine. If the profile is too biased towards the tip or the root of the turbine, it can cause premature failure. The final trimming of this profile can be controlled by the addition of air through the downstream components of the combustor or the high pressure nozzle guide vane platforms.

Military aero engines have an additional requirement to be able to cope with missile plume ingestion. After a missile has been fired, its hot exhaust gases may be ingested into the engine, and the momentary increase in inlet mass flow and temperature, and the

secondary effect of depletion of oxygen must not cause extinction of the flame.

Future trends

In order to meet future emissions requirements, large civil engine combustor design is moving toward a lean-burn approach. This eliminates fuel-rich pockets within the combustor, reducing smoke and NO$_x$ production. It is, however, not without its problems: the weak AFRs within the combustor make the problems of stability, ignition, and relight more difficult. This can be overcome by staging the input of fuel: a 'pilot' fuel supply being used for low power operation and a main supply being brought in for higher power. However, this in turn leads to additional cost, weight, and complexity. The need to switch between two fuel supplies also complicates the control system and fuel system thermal management. In addition, at lean AFRs, slight changes in AFR can lead to large changes in heat release. This can lead to aero-acoustic instability (an audible rumbling sound), which may cause passenger discomfort or fatigue failure of engine components, depending on the frequency of the instability.

Marine and industrial considerations

Marine and industrial gas turbine engines need to contend with different liquid fuels from aero engines. Diesel may have a higher aromatic content than kerosene, which tends

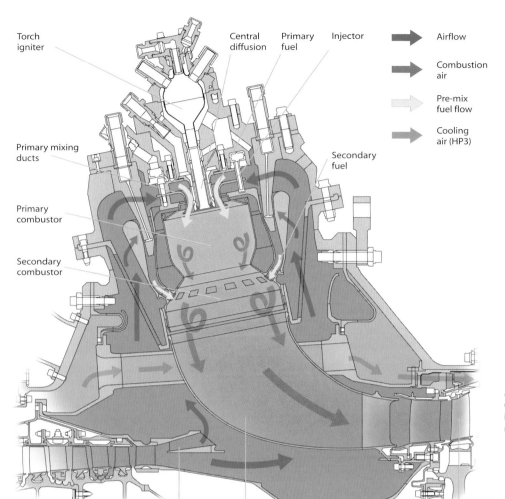

Torch
igniter

Central
diffusion

Primary
fuel

Injector

→ Airflow

→ Combustion
air

→ Pre-mix
fuel flow

→ Cooling
air (HP3)

Secondary
fuel

Primary mixing
ducts

Primary
combustor

Secondary
combustor

Airflow through a DLE combustor as used
on industrial gas turbines. The energy sector
is using new design approaches in order to
reduce environmental impact.

Diffuser

Discharge nozzle

HP turbine

to give an increase in smoke at high power. There is also a tendency for marine engines to produce white smoke when starting; this is due to marine diesel fuels having a higher boiling point than aviation fuels, which causes poor fuel preparation and low combustion efficiency.

Emissions regulations for industrial engines are normally more stringent than those for aero engines, because these engines operate in a fixed location, often near centres of habitation. The regulations are usually set by the end customer to meet local air quality agreements. The low CO and NO_x levels needed may require premixed, lean-burn, staged combustion to maintain a uniform, defined flame temperature across a wide range of power settings and ambient temperatures. Staging may be either series, where fuel is injected at different axial

positions into the same air stream, or parallel, where fuel is injected at different radial or circumferential positions.

Dry, low emissions

The RB211 DLE (dry, low emissions) combustor uses series staging; engine starting uses the conventional, central diffusion flame; at low power, the primary zone is fuelled with premixed gas and air; at high power, the secondary zone is also fuelled. For all such engines, CO and UHC may be further controlled at low power settings by making the engine cycle hotter: air can be bled off or, for fixed-speed compressors, variable inlet guide vanes can be used to reduce the airflow, thereby enriching the fuel/air mixture in the combustor. Alternatively, at high power, water may be injected with the fuel into the combustor, reducing the flame temperature and thereby reducing NO_x production.

Water injection

Water may be introduced up to a water-to-fuel ratio of approximately 1.3:1. After this point, CO and UHC will rise due to reduced combustion efficiency. Smoke will also increase due to quenching of smoke consumption reactions. The introduction of water gives a power boost by increasing the air density but a reduction in cycle efficiency; historically, it was used for many turbojets at take-off, which unlike turbofans have to be sized for take-off.

For industrial engines, the introduction of steam gives both an increase in power and cycle efficiency as some exhaust heat can be recovered. The Pegasus engine makes use of water injection to increase take-off performance, and water injection is being considered for future large civil aero engines at take-off for emissions reduction and life extension.

Gases may leave a modern combustor at temperatures around 1,600°C.
The materials used in the turbine blades melt at 1,200°C.

turbines

THE FIRST TASK OF THE TURBINE SYSTEM IS SURVIVAL. GLOWING RED-HOT, THE BLADES OPERATE IN TEMPERATURES WELL ABOVE THEIR MELTING POINT; EACH BLADE IS BEING STRETCHED BY 18 TONNES OF CENTRIFUGAL FORCE AS IT TRAVELS AT 500 METRES PER SECOND. THE TURBINE'S SECOND TASK IS TO DRIVE THE COMPRESSOR.

turbines

The conventional turbine system is an assembly of alternate static vanes and rotating disc-mounted blades connected to shafts. The blades and vanes are contained in a divergent casing. The turbine produces a rotational power output along a shaft; it usually provides drive to a fan, a compressor and accessories, or, in the case of engines that do not make sole use of a jet for propulsion, it produces shaft power for a propeller, rotor, pump, compressor, or generator. There is a large range of turbine solutions designed and manufactured for civil and military aerospace, marine, industrial, and energy applications.

Improving efficiency through design

Turbine modules are designed, manufactured, and tested in line with the following project criteria:

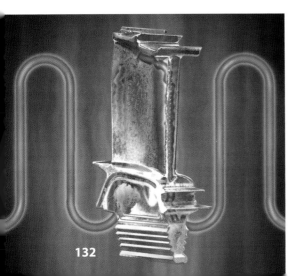

Turbine cooling technology reduces blade temperatures by 400°C; if it could be applied to a blade of ice, this technology would keep the blade frozen indefinitely – even when being 'cooked' at the highest setting of a domestic oven

> providing the required thrust
> minimising cost
> minimising weight
> minimising fuel consumption
> minimising emissions
> minimising delivery timescales.

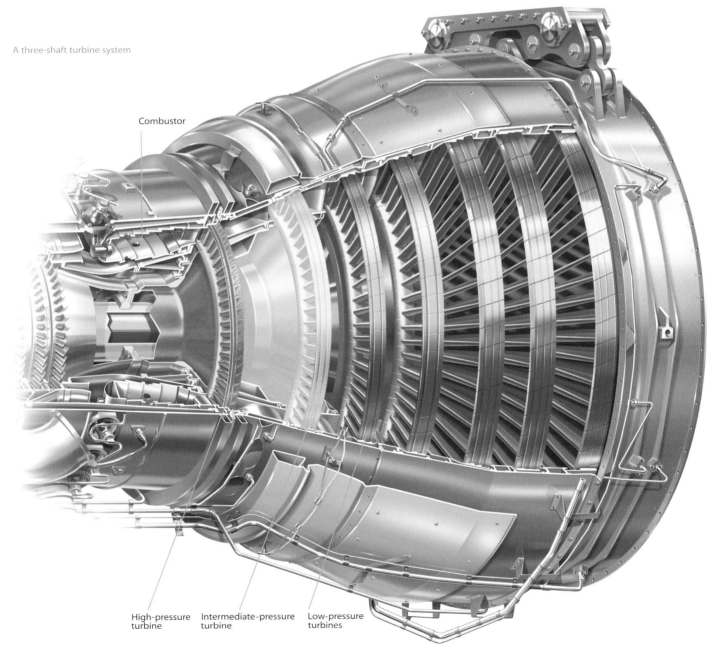

A three-shaft turbine system

Combustor

High-pressure turbine

Intermediate-pressure turbine

Low-pressure turbines

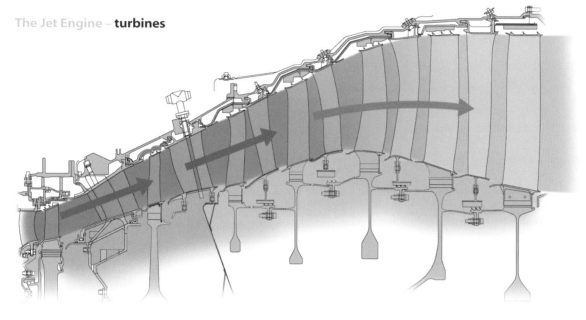

Temperature and
pressure variations
through the turbine
as power is extracted
from the gasflow

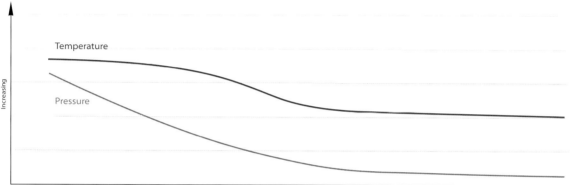

Reducing pressure and temperature through turbines

Basic principles

The turbine assembly is mounted behind,
or downstream of, the combustor, commonly
forming the rear third of a jet engine when
viewed as a whole. Having been highly
compressed, mixed with vaporised fuel,
and ignited, the hot gases leaving the
combustor are expanded to a lower pressure
and temperature through the turbine.
This expansion extracts energy from the gas
to rotate the turbine blades and disc assembly,
which then drives the compressor via
a centrally rotating shaft.

The civil engine market requirements for low
fuel burn and high fuel efficiency are pushing
designs towards engines with a higher bypass
ratio. On turbofan engines, the turbines drive
both a low-pressure compressor or fan
(producing most of the engine's thrust) and
a higher pressure compressor, which ingests
and compresses air ready for the combustion
process. Some turbines drive another
compressor between the low- and high-
pressure compressors. To achieve this,

the air stream is split: some is extracted from
the fan and passed through a duct outside
the turbine and combustor; the remainder
passes through the core of the engine.
To produce the correct driving torque and
efficiency at each stage of the engine, the
turbine may consist of several stages, each
employing one row of static nozzle guide
vanes (NGVs) and one row of rotating blades.
The number of utilised turbine stages depends
upon the relationship between the power
required, the rotational shaft speed, and the
permitted turbine diameter.

As the gas is expanded and work is extracted
from the air passing through each stage of
the turbine, operating temperatures and
pressures reduce accordingly. This means
that the intermediate pressure (IP) turbine
does not need as much, nor as sophisticated,
cooling as the high-pressure (HP) system
– although civil IP turbine and military
low-pressure (LP) components still use
oxidation-resistant nickel alloys to minimise

the required cooling and hence maximise
stage efficiency. Further downstream, civil LP
turbine components can be designed to
be run uncooled, and can be made from
lower temperature capability alloys as the
gas temperature falls to within material
property limits. Turbine exit temperature
from the last LP turbine stage is
approximately 550°C.

Turbine types

There are three types of turbine: impulse,
reaction, and a combination of the two
known as impulse-reaction. In the impulse
type turbine, the pressure drop across each
stage occurs in the fixed NGV, which, because
of its convergent shape, increases the gas
velocity while reducing pressure. The gas
is directed onto the turbine blades, which
experience an impulse force caused by
the impact of the gasflow on the blades.
In the reaction type, the fixed NGVs are
designed to alter the flow direction only,
without changing the pressure.

The converging blade passages experience a reaction force resulting from the expansion and acceleration of the gas. Normally, modern gas turbines rely on a combination of both design styles, and modern aerodynamic design methods enable the characteristics of components to be tailored to maximise work output and stage efficiency.

The mean blade speed of a turbine has considerable effect on the maximum efficiency possible for a given stage output. As rotational speeds increase in the quest for efficiency, so do the forces and stresses involved within the system. Stress in a turbine disc increases as a function of the square of the speed; therefore, to maintain the same stress level at higher speeds, the disc's sectional thickness, and thus its weight, must be increased proportionally. For these reasons, the final design is always a compromise between efficiency and weight. Due to the high proportion of thrust generated by the fan, modern high bypass engines have a better propulsive efficiency than lower bypass ratio designs and so can have a smaller turbine for a given thrust.

A typical civil turbine may have an overall length of up to 1.4m (combining all the turbine stages) and a maximum diameter of up to 1.3m. Military turbines are much smaller, typically under 0.4m in length (across the two stages) with a maximum diameter of about 0.75m. Helicopter turbines are smaller still. In all cases, an increase in turbine rotational speed comes with the reduction in scale in order to optimise work output and efficiency.

The number of shafts and, therefore, the number of turbines can also vary with the type of engine. High compression ratio engines usually have at least two shafts, with two turbines (HP and LP) driving high- and low-pressure compressors. On some high bypass turbofan engines, an IP turbine system is employed between the HP and LP turbines, forming a triple-spool system. In other designs, especially those whose output is shaft power to an external system, driving torque is derived from a free-power turbine. This method allows the free-power turbine to be designed to run at its optimum speed as it is mechanically independent of both the gas generator turbine and compressor shafts.

Power turbines

A power turbine is the means of delivering usable shaft power in an energy or marine application. The power turbine is similar in layout to aero LP turbines and also extracts energy from the hot exhaust gases exiting the gas generator (core of the engine). This energy is converted from an axial gasflow to a rotational mechanical energy by one or more rows of NGVs and rotor blades. The extracted rotational energy is used to drive various pieces of equipment. For energy applications, the driven equipment is usually a compressor, pump, or alternator. For marine, a propeller or an alternator.

The rotational speeds of power turbines vary depending on application; for the smallest engines below 10MW, maintaining blade speed would tend to increase it. A gearbox may be used to match the speed to the requirement of the driven equipment.

Alternators can be designed to run at 3,000rpm (50Hz) or 3,600rpm (60Hz) for electrical generation, which would often be direct drive. Below 30MW, '4 pole' alternators run at 1,800rpm.

For oil and gas pipelines, pumps and compressors typically require speeds between 5,000 and 6,000rpm, and are directly driven. For oil extraction, pump speeds are roughly double this, and a gearbox is used.

Ships use gas generators to drive power turbines in a variety of applications:

❯ In conventional gas turbine-powered ships, there is a mechanical drive from power turbine to propeller via a gearbox.

❯ Recently some ships have adopted electrical drive. Here, the power turbines drive alternators, and electric motors drive the propeller.

❯ Other ships use water jets for propulsion. The power turbine drives a ducted pump. Water is drawn in from beneath the vessel and is ejected at high velocity from the stern of the ship.

There are two general types of power turbines:

❯ Heavyweight – custom designed, high-speed

❯ Aero-derivative – based on the aero engine LP turbine.

A pure impulse turbine compared to an impulse-reaction turbine

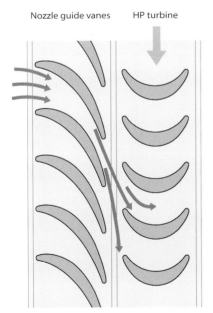

Nozzle guide vanes   HP turbine

Turbine driven by the impulse of the gas flow only

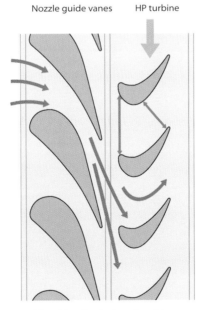

Nozzle guide vanes   HP turbine

Turbine driven by the impulse of the gas flow and its subsequent reaction as it accelerates through the converging blade passage

Typical gas generator and heavyweight power turbine arrangement

The gas generator is not directly coupled to the power turbine other than by ducting to entrain the hot gases. For maintenance, the gas generator is removable, independently of the power turbine. The power turbine may be left in situ at the installation, because the heavyweight construction gives a long life and allows a higher rotational speed than today's aero engine LP turbines. The thrust loads are taken by non-aerospace hydrodynamic bearings, which share a mineral oil lubrication system with the driven equipment.

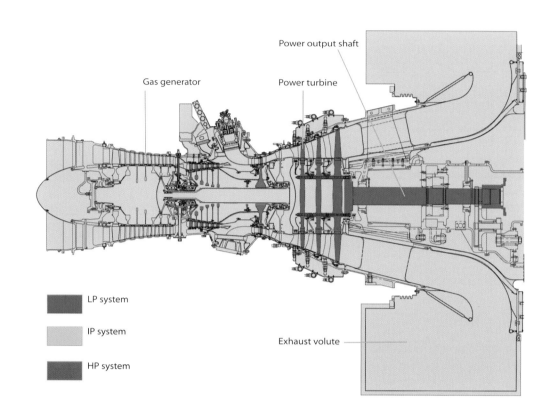

Gas generator

Power output shaft

Power turbine

Exhaust volute

- ■ LP system
- ▢ IP system
- ■ HP system

Typical industrial gas turbine arrangement with aero-derivative LP turbine

In aero-derivative engines, the typical three-shaft construction is retained and the LP turbine – that is, the power turbine – is contained within the gas turbine. The LP turbine rotational speed typically matches a driven alternator. For maintenance, the whole gas turbine is removed.

Similarly, there are two general concepts of marine power turbines. In the heavyweight approach, the power turbine is normally installed for the life of the ship.

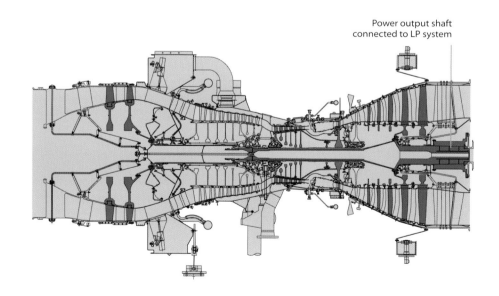

Power output shaft connected to LP system

CFD analysis of HP turbines

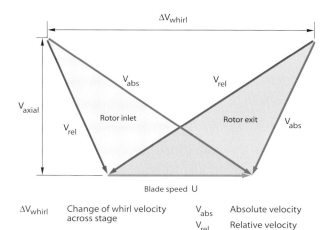

A turbine blade triangle shows the gas path angles into and out of the blade. The change in swirl velocity and blade speed is used with the mass flow rate to calculate turbine power output.

ΔV_{whirl} Change of whirl velocity across stage

V_{abs} Absolute velocity

V_{rel} Relative velocity

Turbine design methodology

Many requirements must be met when a new turbine aerodynamic design is prepared. The three main aerodynamic objectives are to produce sufficient turbine power, to pass the correct amount of gasflow, and to achieve (or better) target stage efficiency. Complex 3D aerodynamic designs are used to accurately tailor the aerodynamic shape of NGV and turbine blade aerofoils – and platforms – to suit the required stage characteristics. The flow characteristics of the turbine must be carefully matched with those of the compressor to achieve efficiency and performance targets. If the turbine components allowed too low a maximum flow, then a back pressure would build up in the engine causing the compressor to surge. Conversely, too high a flow would cause the compressor to choke, where the total gasflow entering the compressor is greater than its working capacity due to the imbalance between the two systems. Either condition would induce a loss in engine efficiency and performance. Modern aerodynamic design not only meets this criterion but also incorporates features to minimise both boundary layer flow losses and also the NGV wake forcing effects on rotors. Every effort is made to minimise the effects of consumption, and reintroduction, of cooling air into the gas path.

Every design is a compromise and the design methodologies used often require a lengthy iteration process to achieve the best overall solution. This series of iterative loops is required because of each component's inter-relationship with its neighbouring components. For example, a modified blade design may necessitate a redesign on the shroud or a change in the hub. Any change in the blade may also dictate a change in the disc design. A disc iteration may

then affect the containment requirements, possibly affecting the casing design criteria – and so on.

A new turbine component will be reviewed by the following disciplines before engine development testing begins:

› aerodynamic design

› cooling or thermal design and analysis

› stress analysis

› mechanical design

› manufacturing.

The component's operation is then fully proven and validated before certification is received from the relevant authority and the product is rolled out.

Energy transfer from gas flow to turbine

The turbine power output (to the compressor or load) depends on the effective transfer of energy between the expanding combustion gases and the turbine stator and rotor. The amount of power developed by each blade is proportional to the gas mass flow rate, blade speed, and change in swirl velocity of the gas.

The energy transfer between the working fluid and the turbine does not achieve 100 per cent efficiency due to thermodynamic and mechanical losses. These inefficiencies include aerodynamic losses across the NGV and blade, overtip leakage losses with the rotor, the efficiency deficit effects through the use of compressed cooling air, and the leakage of cooling air between adjacent components. Modern turbines operate at levels of efficiency

greater than 90 per cent; this is only achieved through careful iteration and design optimisation. Shrouded military turbines tend to achieve similar levels of stage efficiency, but overall output efficiencies are reduced due to the lower bypass ratios on smaller military designs.

Following the combustion process, the gas is forced through the combustor discharge nozzles into the HP NGVs where, because of their aerodynamic convergent shape, it is accelerated to about the speed of sound (about 850m/s at a high turbine entry temperature). Simultaneously, the gas is swirled in the direction of the turbine blades' rotation. As the tailored gas flow enters and passes through the turbine blades and energy is extracted, their aerodynamic form creates torque, a rotational reaction force across each blade, causing them to turn the disc and shaft assembly, driving the compressor.

The torque or turning power generated by the turbine is governed by the mass flow rate and the energy transfer between the inlet and the outlet of the turbine blades. The design of the turbine is such that the swirl of the gasflow will be removed by its operation, and so the flow at the exit of the turbine will be substantially more axial as it flows into the exhaust system. Excessive residual swirl reduces the efficiency of the exhaust system and can also produce jet pipe vibration, affecting strut and exhaust support integrity. This also explains why each stage of a conventional turbine requires an NGV to recondition the flow with appropriate swirl and axial velocity for the receiving downstream rotor.

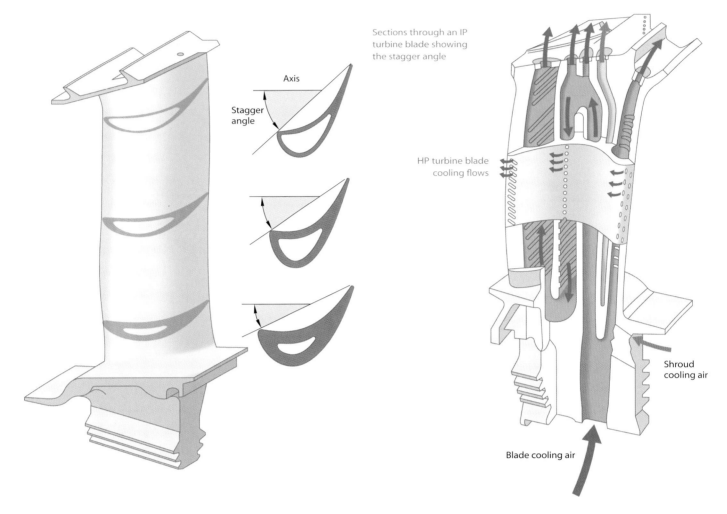

Sections through an IP turbine blade showing the stagger angle

Axis

Stagger angle

HP turbine blade cooling flows

Shroud cooling air

Blade cooling air

Statorless turbines have been designed, using the latest aerodynamic design methods. The upstream rotor exit velocities and remaining swirl are tailored to suit the inlet requirements of the following rotor, which will counter-rotate to maintain efficiency. The targeted benefits of such designs include weight reduction, minimised engine length, and a significant reduction in the total number of components used. However, balancing the work between turbine stages is the on-going challenge for these designs. Similarly, in an effort to improve efficiency, contra-rotating turbine designs have been tested with promising results, particularly in military turbines. Here, the HP turbine rotates counter to the IP (on three-shaft engines) or LP turbine, and enables the designer to tailor the exit velocities and vectors from one stage to the next. On three-shaft engines, this frees the aerodynamic design, allowing gains in IP NGV aerofoil performance and hence stage efficiency to be improved. In military engines, significant aerodynamic and mechanical design improvements have been achieved through the use of contra-rotating stages.

It is evident when viewing turbine blade and NGV designs that the aerofoils are twisted along their length, with a greater stagger angle at the tip than at the root of the aerofoil. This ensures the gasflow from the combustor is optimised along the component's entire height (span), and so the flow continues downstream of the rotor with uniform axial velocity. The magnitude of rotational force varies from root to tip, being least at the root and highest at the tip, with the mean value at approximately 50 per cent span.

Turbine cooling
Working environment
At approximately 1600°C, HP turbine components in the hottest part of the gas stream are designed to operate five times hotter than a typical domestic oven. These temperatures are far greater than the melting point of the leading nickel-based alloys from which they are cast.

The HP blades, NGVs, and seal segments are therefore cooled internally and externally using cooling air from the exit of the HP compressor, itself at temperatures over 700°C (achieved through compression only) and fed at a pressure of 3,800kPa. The gas stream pressure at turbine inlet is over 3,600kPa; therefore, the cooling feed pressure margin is only small and maintaining this pressure margin is critical to component operation.

Considerations in deciding whether a blade or vane should be cooled or uncooled include the choice of materials, the use of a thermal barrier coating (TBC), the performance requirements, and the engine cost target. Not cooling a blade or vane gives more freedom in terms of aerofoil design, both size and shape, as no internal cooling system has to be cast within it. It will, however, limit the component's operating temperatures, affecting performance, while also limiting the scope for future engine growth. TBCs alone provide no benefit in reducing metal temperatures on uncooled turbine components. An uncooled component may also have to be manufactured from an improved material, affecting cost and manufacturability.

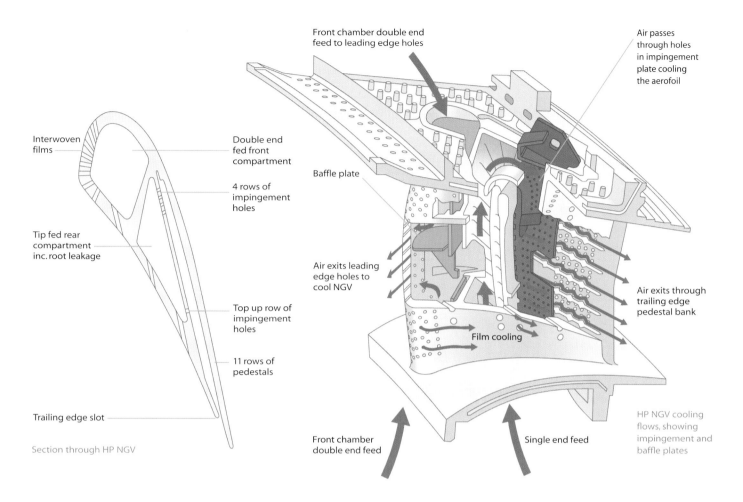

Section through HP NGV

- Interwoven films
- Double end fed front compartment
- 4 rows of impingement holes
- Tip fed rear compartment inc. root leakage
- Top up row of impingement holes
- 11 rows of pedestals
- Trailing edge slot

Front chamber double end feed to leading edge holes

Air passes through holes in impingement plate cooling the aerofoil

Baffle plate

Air exits leading edge holes to cool NGV

Air exits through trailing edge pedestal bank

Film cooling

Front chamber double end feed

Single end feed

HP NGV cooling flows, showing impingement and baffle plates

Advances in metallurgy and casting technology have enabled the use of single crystal nickel alloy components. The resulting improvements in material properties allow the components to be run at increased turbine operating temperatures. The use of advanced alloys cast in this way improves life limits by enabling the most efficient use of cooling air and by giving the designer a better understanding of the material properties.

Nickel alloys are an almost universal solution for high temperature turbine blades and NGVs – due to their high temperature creep resistance, and strength retention. Single crystal components have superior metallurgical properties in all directions, but come at a far greater manufacturing cost. Similar alloys can be cast utilising the directionally solidified method, which is cheaper than single crystal for a small reduction in properties, or as a

conventional Equiax casting, further reducing cost and material limits. Overall, the turbine design and materials selection is dependent on the trade balance between temperature, life, and component cost.

Cooling geometry design itself has improved significantly over the years, with patented laser-drilled cooling hole designs and soluble ceramic core technologies enabling

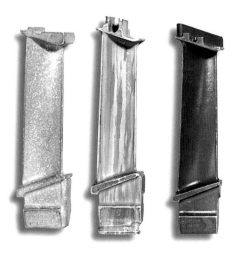

Three common turbine blade casting options balancing cost, yield, and performance: Equiax, directionally solidified, and single crystal alloys

Comparison of turbine blade life properties

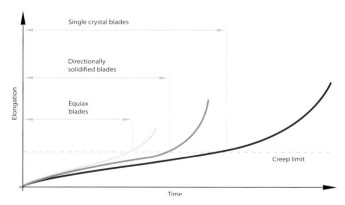

- Single crystal blades
- Directionally solidified blades
- Equiax blades
- Creep limit

Elongation

Time

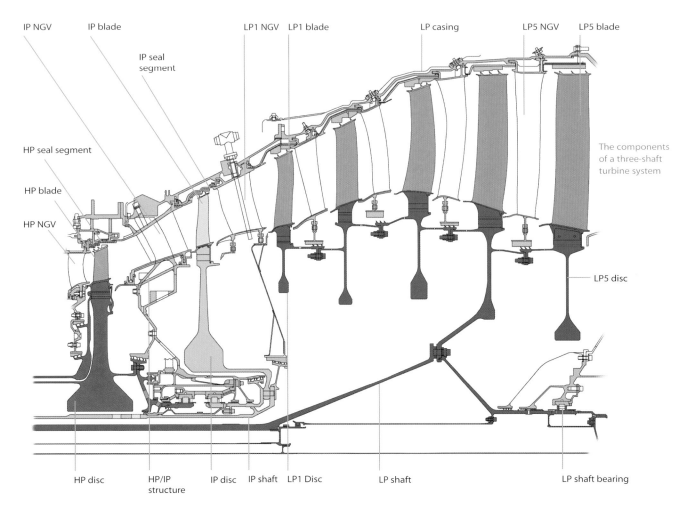

IP NGV   IP blade   LP1 NGV LP1 blade   LP casing   LP5 NGV LP5 blade

IP seal
segment

HP seal segment

HP blade

HP NGV

The components
of a three-shaft
turbine system

LP5 disc

HP disc   HP/IP   IP disc IP shaft LP1 Disc   LP shaft   LP shaft bearing
           structure

enhanced cooling methods with high levels of cooling effectiveness on blades and vanes. These methods enable the reduction of cooling airflow – as does the controlled application of ceramic TBCs.

Cooled components allow higher turbine operating temperatures, producing increased thrust levels. Again, the decision is formed through compromise; if a component is to be cooled, it is also necessary to balance the amount of cooling flow and the cooling design effectiveness. High cooling flow designs mean that excessive compressor air is bled away from the core flow prior to combustion. This impacts on turbine performance in two distinct ways: first, a cycle penalty is incurred through not combusting the volume of air used for cooling the turbine components, reducing the amount of energy transferable in the turbine; secondly, aerodynamic losses are induced by re-introducing this air through cooling holes from the components and out into the gas path. Designs with the most effective cooling can often increase complexity of manufacture and therefore component cost.

Turbine components

All turbine components are designed in line with stringent design rules and requirements set by the customer on performance, cost, weight, life, and timescale. A typical turbine assembly can be broken down into five main component types: casings and structures, discs, shafts, NGVs, and blades.

Casings and structures

The casings form the outer structure of the turbine and enclose the hot gases exiting the combustor. They are normally constructed from forged steel or nickel alloys that must be strong enough to contain the internal gas pressures of the turbine. The casing must also contain any debris if a component fails. Turbine casings are designed to transmit and react the axial and torsional loads imposed by the turbine assembly.

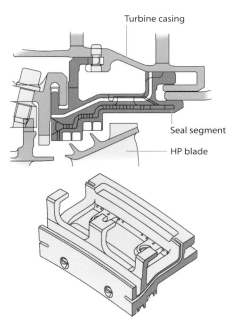

Turbine casing

Seal segment

HP blade

HP turbine blade seal segment

Structures are designed to connect these casings to the internal shaft bearing supports, transmitting the bearing loads into the case and stiffening the assembly (》 152 – 154). Air and oil systems, required to lubricate and cool the bearings, may pass through the casing and structures.

Other static component types fit into the casings to form the completed assembly, including NGVs, seals (such as segments to seal a rotor path), and supporting rings. These components are retained in the casing by a variety of methods including dowels, hooks, and anti-rotation features. Seal segments typically form a peripheral ring of abradable material around the blades' rotating tips. In some cases, the rotors' tip fins cut circumferential grooves into these components' softer, abradable honeycomb material, forming a controlled labyrinth air seal and minimising leakage over the rotors' tip fins. It is essential to control the thermal movement of the seals so that optimum blade tip running clearance is maintained.

The thermal expansion of the casing can be controlled throughout the engine cycle by using compressed cooling air to maintain optimum clearances between blade tip and seal. The cooling air is fed from the compressor into the case-mounted cooling manifolds. This effectively minimises the blade overtip leakages and helps maximise stage efficiency. Such a method of tip clearance control can be either active or passive and both can be controlled by modulated cooling airflow. Tip clearance control allows higher turbine temperatures and shaft speeds to be used, especially when used with shroudless turbine blades.

In order to support engine health monitoring (》 199, 261-262), inspection of gas path components using borescopes must be catered for. This requires access ports to be provided within the casing.

Instrumentation such as thermocouples forming part of the engine control system may pass through the casings into the static components within.

Discs

The main function of the turbine discs is to locate and retain the rotating blades enabling the circumferential force produced by them to be transmitted to the compressor through the central shafts. Each row of blades is retained in the rim of a disc, via a root fixing – commonly of fir-tree design – designed to withstand the enormous centrifugal loads exerted onto the disc by the mass of the blades rotating at high speed. The disc has drive arms connected to a corresponding stage of the compressor via a shaft.

Discs are typically formed from nickel alloy forgings, the raw materials for which are carefully selected and inspected for lack of defects prior to and during disc manufacture. Alloys have been specifically developed for high-strength disc applications. Modern alloys and powder metallurgy have produced an increase in strength allowing faster shaft speeds or higher temperatures to be achieved.

Discs are classified as critical parts (that is, any part whose failure has hazardous effects on engine, airframe, ship, or installation). The risk of disc failure is mitigated through careful material selection and adherence to strict design criteria. Design criteria on ultimate tensile stress, proof stress, creep, and fatigue all have to be satisfied.

Shafts

The turbine shafts have three main functions: transmitting torque from the turbine to the compressor, transmitting axial loads to the compressor and location bearings, and supporting the disc and blade assemblies. The turbine shafts are carried on oil-cooled and lubricated bearings mounted within the structure; they may be common to the compressor shaft or connected to it by a self-aligning helical spline coupling. On a modern civil three-spool engine the three shafts each rotate concentrically within one another at their own optimum speed. Typically, at take-off condition the LP shaft rotates at 3,000rpm, the IP shaft at 6,000rpm, and the HP drum at 10,000rpm. Military designs tend to incorporate two shafts only, with HP and LP turbines rotating at their optimum speeds, typically much faster than larger civil engines due to their smaller diameters.

Nozzle guide vanes

NGVs are designed to convert part of the gasflow's heat and pressure energy into a tailored kinetic energy from which the rotor blades can generate power. They are shaped to swirl the gasflow in the direction of the rotor's rotation, maximising rotor efficiency. In doing this, the tangential momentum of the gas is increased.

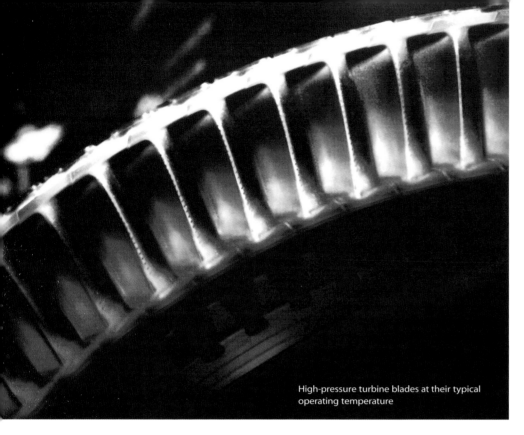

High-pressure turbine blades at their typical operating temperature

into the vane aerofoils (and sometimes the vane platforms) at a higher pressure than that of the surrounding gas path. This pressure differential flows the cooling air through rows of machined film cooling holes, bathing the components gas-washed exterior in a film of cool air. Without this film flowing onto and over the gas-washed surfaces, the vane temperature would quickly exceed the melting point of the alloy. To minimise the amount of cooling air required by the component, modern civil HP and IP NGVs are cast using single crystal nickel alloys and are typically coated in a ceramic TBC. This significantly reduces the component thermal conduction and therefore its external metal temperature.

Blades

The turbine blade is designed to generate power by translating circumferential aerodynamic forces on the aerofoil to the rotating disc. The blades are of an advanced aerofoil shape, designed to provide passages between adjacent blades that give a steady acceleration of the flow up to the throat, where the area is smallest and the velocity reaches that required at exit to produce the necessary degree of rotation.

The blades rotate within the casings with a typical tip speed of 460m/s. At this speed, the power output of a single civil HP blade is ten times higher than a small family car and the force transmitted into the disc by each blade at red line speed is approx 18 tonnes – that is a g-force of 66,000g.

NGVs are static components (sometimes referred to as stators), mounted into the turbine casings, designed both to withstand the axial and torque loads imparted from the gas stream and to react thermally without inducing high internal stresses within the assembly. They are located using machined hooks or rails, fixed with pins and dowels, and react circumferentially against casing-mounted anti-rotation features. The NGVs are designed to articulate with relative thermal movement between their casing mounts, while maintaining effective air seals to protect the gas path and cooling air system from leakage.

The assembled NGV gas path results in a set of individual windows forming an aerodynamic 'throat'. Designed to achieve optimum stage efficiency and for compatibility with compressor and combustion design, modern NGVs are of an increasingly complex curved aerofoil shape.

NGVs in modern civil HP and IP turbines tend to be cooled; LP NGVs are often run uncooled. Military designs use cooling in both the HP and LP turbine stators. Internally cooled components are manufactured by investment casting with complex core geometries, maximising the cooling effectiveness of the compressed air in use. Cooling air is flowed

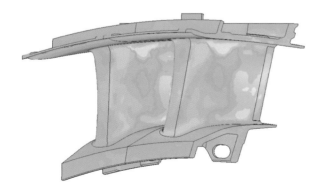

Thermal stress analysis of a nozzle guide vane assembly

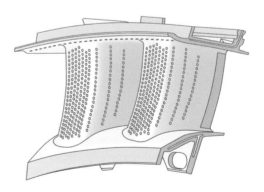

HP nozzle guide vane assembly

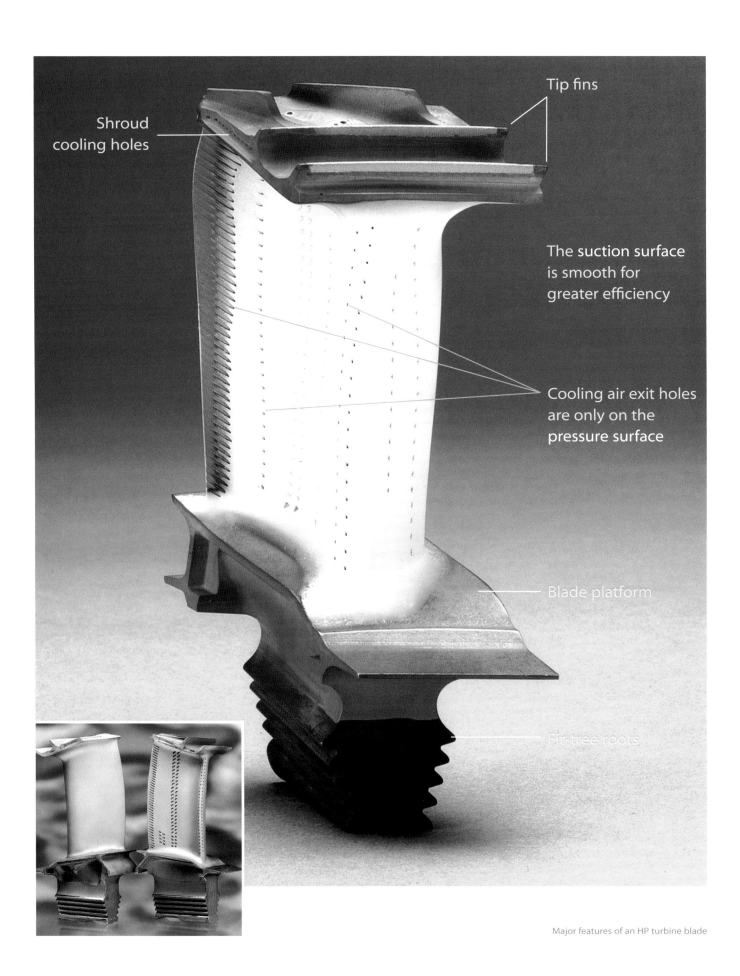

Tip fins

Shroud
cooling holes

The **suction surface**
is smooth for
greater efficiency

Cooling air exit holes
are only on the
pressure surface

Blade platform

Fir-tree roots

Major features of an HP turbine blade

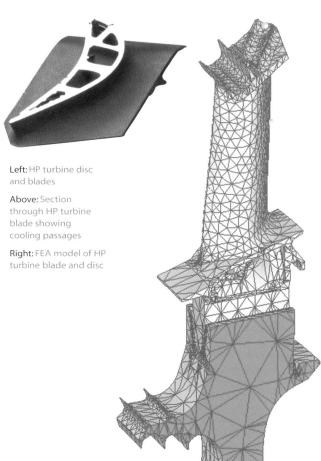

Left: HP turbine disc and blades

Above: Section through HP turbine blade showing cooling passages

Right: FEA model of HP turbine blade and disc

The blade's cross-section design is governed by the permitted stress in the material used and by the size of any core passages required for cooling purposes. The hottest running blades are cast in a high-temperature nickel alloy using the lost wax casting method and are often coated in a ceramic TBC on their aerofoils and platforms. As with the NGVs, operating temperatures dictate the need to internally cool the HP blades with cooling air flowing through a complex internal channel system before exiting through rows of cooling holes. Cooling flow is detrimental to turbine performance, and is regulated very carefully – and therefore, blade material selection is very important. The blades glow red-hot during engine running, yet at this condition they must still be strong enough to carry high centrifugal loads due to their rotation and the bending load due to the gas stream. They must also be resistant to fatigue, thermal shock, corrosion, and oxidation.

Blades may incorporate a shroud at the tip, forming an outer annulus ring when assembled. Shroudless blades can be run at higher rotational speeds due to their lower mass but suffer from a potential increase in overtip leakage and resultant performance effects. The blades are commonly mounted into the disc by fir-tree fittings, designed and carefully machined to distribute the running loads equally between each serration.

Over a period of operational time, the turbine blades slowly increase in length – this phenomenon is known as creep. Creep life and material oxidation limits will dictate the finite useful life limit of the component. Typically, a modern civil blade will be designed to work under operational conditions for 35,000 hours before it is overhauled or replaced. On an airliner that flies 14 hours a day, that is 6 years on the wing and 15 million flight miles between major services.

Evolving design considerations

As turbine design progresses through each new engine project, it is important to remember that the basic design and operating principles remain the same as those used in the very earliest of turbine designs. Today, modern market requirements combined with reduced timescales add pressure to the design and development programme. The focus of investment and development on the latest products is channelled towards ever more demanding targets in turbine performance and efficiency – together with reductions in fuel burn, unit cost, and engine weight.

This development relies heavily on improvements in material properties, allowing increased turbine operating temperatures with less compressor cooling air, while increasing speeds (and therefore component load) to achieve the advancing design intent.

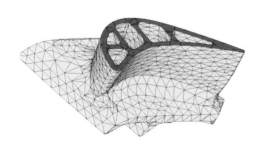

FEA model of a HP turbine blade tip

The development of next generation materials, often for specific functions within the turbine system, is very important, allowing large steps forward in thrust and efficiencies. In parallel, the latest cooling design geometries enable improvements in the effective use of cooling air. With reductions in flow combining with the latest high temperature materials and improved TBC systems, the performance of turbine components is proving more efficient with each iteration. Weight and cost reduction initiatives are also paramount in design, particularly through manufacturing improvements where refined methods and modern technologies are employed to minimise unit costs.

Turbine components are now designed from concept with ease of manufacture and assembly in mind – by taking the lessons learnt from the previous designs. In this way, Design For Manufacture (DFM), and Design For Assembly (DFA) have become key to the development of a fit-for-purpose, cost effective design solution. The cost of ownership is also considered, with a significant effort on latest designs being aimed at aftermarket, overhaul, and repair requirements.

The gases flowing through the turbine transfer energy to the rotors.
This energy must now be put to work.

transmissions

TRANSMISSION: THE TRANSFERENCE OF MOTIVE FORCE
– POWER – FROM ONE COMPONENT TO ANOTHER.
THE TRANSMISSION SYSTEM HAS A DUAL ROLE:
TO TRANSFER MECHANICAL POWER WHEREVER IT IS
NEEDED, WHILE STILL RETAINING THE ROTOR SYSTEMS
WITHIN THE ENGINE. NEITHER IS AN EASY TASK.

transmissions

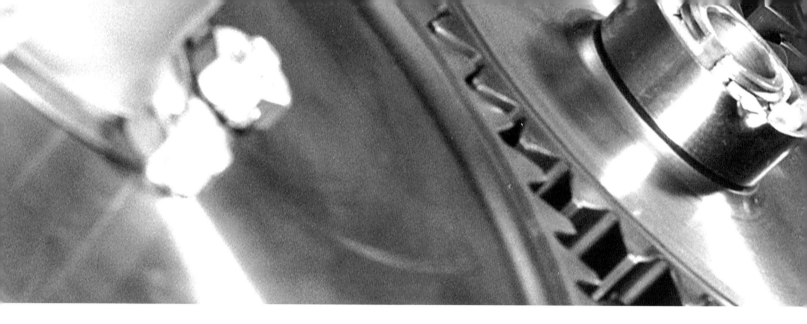

Power transmission is carried out by four component groups:

❭ rotor support structures

❭ gearboxes

❭ shafts

❭ bearings.

These component groups are very different in form, but have two functions in common:

❭ the transfer of mechanical power

❭ the support and location of other engine components.

Rotor support structures

Rotor support structures are large, strong, weight-efficient castings or fabrications that support the engine rotors while allowing the primary engine airflow to pass through.

Gearboxes

On jet engines, the gearboxes provide mechanical power to engine-driven accessories The external gearbox provides a mount for the accessories and distributes mechanical power to, or from, each accessory unit.

Shafts

Shafts transmit power both from the turbine to the compressor, which can be of the order of 75MW (100,000hp), and between internal bevel gears and external gearbox.

Bearings

Provide axial and radial positioning of rotating components; roller bearings provide radial positioning only.

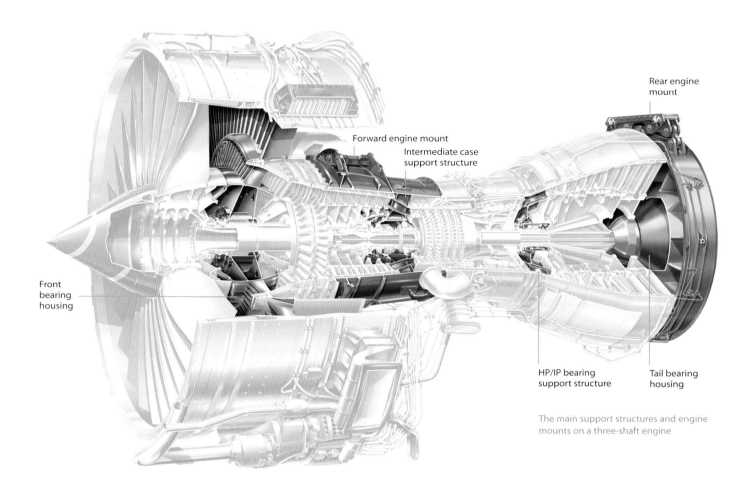

Rear engine mount

Forward engine mount

Intermediate case support structure

Front bearing housing

HP/IP bearing support structure

Tail bearing housing

The main support structures and engine mounts on a three-shaft engine

Rotor support structures

Fundamentally, the engine outer structure is a pressure vessel that contains hot, flowing air. The rotor support structures extend inside the pressure vessel to support the rotating components of the engine while allowing air to pass through from front to rear. They are generally circular, with a number of struts or vanes joining the inner and outer rings and a bearing housing located in the middle. Inside the bearing housings, the bearings allow free rotation, yet precise centring, of the rotors. On the outside, support structures may provide mounting lugs as attachment points for external engine components or the engine-to-aircraft mounting. Some lugs also transmit engine thrust loads to restrain forward and reverse motion. The support structures are joined together by compressor or turbine casings to form a complete support frame for the engine.

Each engine rotor requires two or even three rotor support structures; however, a single support structure like the intercase can be used for up to three rotors. Because of this, the RB211 and Trent engines need only four structures to support the three rotor systems.

The engine rotors transmit loads generated by the rotors to the stationary engine structure through the rotor support structures. The outer engine structure collects the loads and transfers them to the aircraft at the engine mounts. When an aircraft performs manoeuvres, the structures maintain the centring of the engine rotors. In the event of a component failure, the structures ensure that the engine will not create a hazard to the aircraft, although the engine may stop operating.

The rotor loads enter the support structures at the bearings, which are inside an annulus of flowing air. Struts or vanes transmit the loads through the gaspath to the outer stationary structure of the engine. Both struts and vanes minimise disruption and pressure loss in the gaspath, but vanes are more

sophisticated, and are used to significantly redirect the air/gas. The struts or vanes also provide a path for lubricating oil to be provided to, and returned from, the bearing chambers.

Each structure must withstand a wide range of extreme conditions to ensure the engine's safe and reliable operation. In an aircraft engine, weight must be stringently controlled. Jet engines for aircraft propulsion use the lightest possible materials. In cooler locations, light alloys such as aluminium or magnesium perform well. For moderate temperatures, titanium, though expensive, provides the necessary qualities of high strength, low weight, and temperature capability. In the very hottest locations, heavy materials such as nickel alloys provide sufficient temperature resistance. To ensure the most structurally efficient configurations, engineers use extensive Finite Element Analysis to evaluate the ability of the structures to withstand engine and aircraft loads.

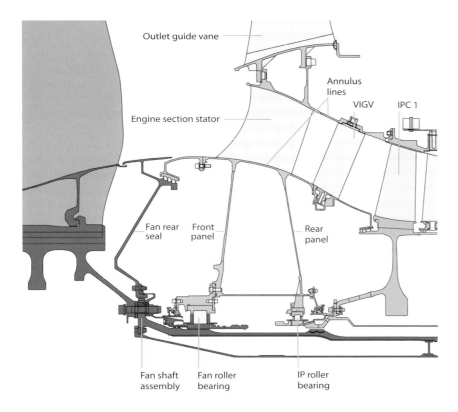

Outlet guide vane

Engine section stator

Annulus lines

VIGV

IPC 1

Fan rear seal

Front panel

Rear panel

Fan shaft assembly

Fan roller bearing

IP roller bearing

The intercase: on a three-shaft engine, it locates the thrust bearings for all three rotors

The innermost part of the support structure is the bearing chamber, which provides a favourable environment for the bearings. Inside, oil nozzles distribute lubrication to the bearings and gears. Around the shafts, labyrinth seals prevent the oil from leaking out and limit the amount of hot air entering the chamber. Buffer air at higher pressure surrounds the bearing chambers outside the labyrinth seals so that air flows inward through the seals. This inward flow of buffer air prevents oil from migrating out of the labyrinth seal.

As described earlier, the RB211 and Trent families use four rotor support structures:

› the front bearing housing

› the intermediate case

› the HP/IP structure

› the tail bearing housing.

Front bearing housing
The front bearing housing (FBH) provides support near the front of the fan rotor (also known as the low-pressure or LP rotor) and at the front of the intermediate-pressure (IP) rotor. The bearing chamber on the inside contains the forward LP and IP roller bearings. The engine section stator (ESS) vanes direct

a portion of the fan airflow into the core of the engine, and carry structural loads to the splitter area. Two conical panels attach the ESS vanes to the bearing chamber.

The ESS structure is manufactured as either a machined cast ring of vanes, or built up from individual forgings that are welded together to form a ring. In addition to structure, the ESS ring provides:

› aerodynamic functionality to feed the IP compressor, delivered by the ESS aerofoil shape

› routing for services, which can include oil feed, oil scavenge and oil and air vent, and speed probe wires.

Intermediate case
Also called the intercase, this is a structure between the HP and IP compressor cases, which houses the main shaft thrust bearings and carries the rotor gas loads through struts to the engine casing and thrust mounts. It also houses the internal gearbox, which incorporates a bevel-gear drive shaft linking the HP rotor to the external gearbox.

The intercase provides support for all three rotor systems. Thrust bearings contained in

the intercase bearing chamber, away from the hot end of the engine, provide mid-rotor support for the LP and IP rotors, and forward support for the HP rotor. These bearings transmit all of the axial forces of the rotors to the engine structure. On Trent and RB211 engines, lugs on the intercase transmit the engine thrust to the nacelle structure. Therefore, unlike the other rotor support structures, the intercase must be strong in the axial direction as well as the radial direction. The intercase, therefore, literally pulls the aircraft through the air.

Because of the location of the intercase at the forward end of the HP rotor, it is called upon for an additional and unique function. It provides an internal gearbox in the bearing chamber to transmit power to and from the HP rotor. This is necessary for engine starting, and to drive mechanical units such as oil pumps and generators that are mounted on the engine. The internal gearbox includes a pair of bevel gears mounted within the intercase. One gear is mounted on the HP rotor; its mating gear is connected to a small shaft that runs through a strut. This small shaft is the radial drive, which is part of the system that transmits mechanical power to and from the external gearbox.

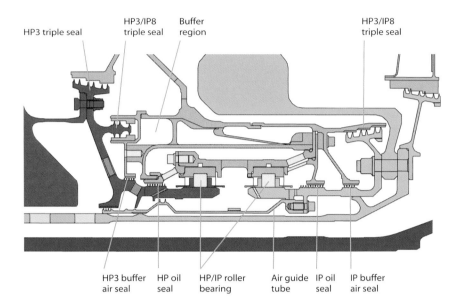

HP3 triple seal
HP3/IP8 triple seal
Buffer region
HP3/IP8 triple seal

HP3 buffer air seal
HP oil seal
HP/IP roller bearing
Air guide tube
IP oil seal
IP buffer air seal

HP/IP hub structure on a three-shaft engine. This structure houses the HP and rear IP roller bearings.

As in all rotor support structures, the intercase contains a passage for the engine's core airflow. This passage is known as a 'swan neck' duct, because is sweeps from the larger radius of the IP compressor exit to the smaller radius of the HP compressor inlet, giving the appearance of a swan neck on drawings. Struts to carry structural loads across the flowpath span the swan neck. The hollow struts in the swan neck duct allow oil services to, and venting from, the bearing chamber, as well as a place for the radial drive.

The structure is usually cast titanium with some amount of welding necessary due to the complexity of the structure.

HP/IP structure

The HP/IP turbine bearing support structure is located between the HP and IP turbine discs to provide support to the aft end of the HP and IP rotors. The bearing chamber houses the HP and aft IP roller bearings. The structure transmits radial bearing loads through the hub and into the outer casing.

This structure operates in a very challenging environment. It is surrounded by very hot engine parts, and must carry load through the HP turbine exit airflow, which is one of the hottest parts of the engine. Struts connect the inner structure to the turbine case while allowing air to pass. In this environment, the only fluid available for cooling is the oil supply, which must pass through the hot flowpath along with the struts.

Due to their interaction with the outer casings, the bearing support structure has a major influence on the control of blade tip clearances and shaft dynamics. The bearing support structure, therefore, must have sufficient stiffness to withstand extreme manoeuvres, while maintaining an adequate fatigue life.

The blade tip clearances are further influenced by the use of an oil squeeze film damper (>> 161). The damper consists of a narrow oil-filled gap between the bearing outer race and the bearing chamber structure. The rotor system, though precisely balanced, will still have unbalance present. The damper provides fluid support for the bearing race in a way that allows the rotor system to rotate about its true mass centre. In addition, the fluid film reduces the unbalance forces transmitted to the structure.

The bearing chamber components operate near the maximum permissible limits for bearings and engine oil. Typically, nickel alloys are the materials used to make the housings, shafts, support structures, and air seals for high bypass three-shaft engines.

Tail bearing housing

The tail bearing housing is the bearing chamber that supports the aft end of the LP rotor, and contains the rear engine mounts. Exit guide vanes provide structural support of the bearing chamber, and provide the pathways for oil, air, and instrumentation

cables. The exit vane shape is simpler than the front bearing housing ESS vanes because they need to provide less turning of the airflow.

The rear-bearing chamber provides a protected environment, housing the LP shaft rear roller bearing and LP turbine over-speed probe. Roller bearings transfer radial loads into the structure. Oil transfer routes through the exit vanes provide oil lubrication, scavenging, and an oil film damper.

Although the tail bearing housing must operate in the environment of the LP turbine exhaust, it is not as severe as that endured by the HP/IP support. Due to the prevailing high temperatures from the turbine, the structure material is a nickel alloy. In the quest for lower production costs, manufacturing methods have varied between a fully cast structure to a fabricated structure, but both methods have proven comparable. The bearing chamber housing is traditionally manufactured from cast steel alloy, but has also been made from cast nickel alloy.

Gearboxes

The jet engine is called upon to provide mechanical power to a number of accessories. These accessories may have a strictly engine-related function, or may provide services to the aircraft. Typical engine accessories include starter, fuel pump, oil pump, alternator, and breather. Typical aircraft accessories include generators and hydraulic pumps. The high level of dependence upon these units requires an extremely reliable drive system that transfers power from the innermost part of the engine, the internal gearbox, to the outermost, the accessory gearbox mounted on the fan case.

An accessory drive system on a three-shaft engine takes between 400 and 500hp from the engine.

Internal gearbox

The need to start the engine by rotating the HP rotor dictates the location of the internal gearbox within the core of an engine. Theoretically, any of the rotors can be used to power the accessories. Historically, the simplest solution has been to use the starting bevel gears to extract power as

well as to provide cranking for starting. Taking power from one rotor while starting a different rotor introduces the need for additional bevel gears and their associated complexity. However, extracting accessory power from the IP or even the LP rotor introduces a number of advantages including reduced fuel consumption and improved engine operability.

Oil nozzles supply oil for lubrication and cooling of the internal gearbox bevel gears and bearings. The internal gearbox is tightly packed with high-speed, rotating components. Therefore, effective scavenging of the spent oil is important in order to minimise windage-driven power loss and associated oil heating. Dedicated scavenge pump elements suck the spent oil from strategic spots in the gearbox.

In most cases, a large amount of air passes through the internal gearbox vent system, thereby removing entrained oil. This aids in the scavenging process.

The radial drive shaft transmits power from the internal gearbox to the accessory gearbox. It also serves to transmit the high torque from the starter to rotate the HP system for engine starting. The radial drive is as slender as possible to fit through a strut with the minimum possible disruption of the airflow. Generally, within a three-shaft engine, the drive shaft locates within one of the intercase struts (normally around bottom dead centre).

Intermediate gearbox

The requirement for an intermediate gearbox (commonly referred to as a step-aside gearbox) is primarily driven by the remote location of the accessory gearbox relative to the internal gearbox. Without an intermediate gearbox, torque transmission from the HP compressor to the accessory gearbox on the fan case would require a single shaft so long that it would impose an impractical whirl

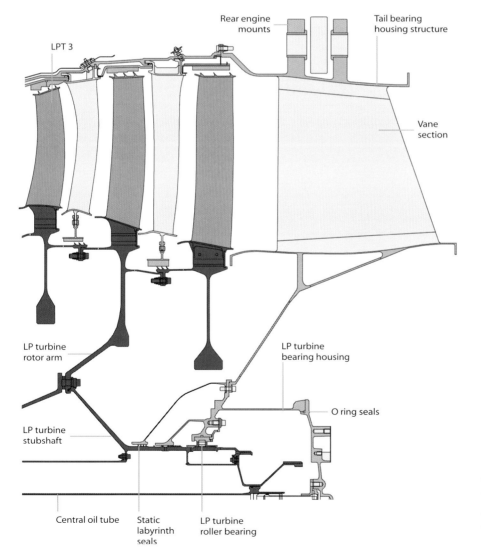

Rear engine mounts

Tail bearing housing structure

LPT 3

Vane section

LP turbine rotor arm

LP turbine bearing housing

O ring seals

LP turbine stubshaft

Central oil tube

Static labyrinth seals

LP turbine roller bearing

On a three-shaft engine, the rear engine mounts are part of the tail bearing housing, which also locates the rear LP roller bearing

margin – known as a supercritical shaft. Instead, the intermediate gearbox, which usually mounts on the compressor casing, provides an intermediate earthing point that permits the use of the short, high-speed, radial drive shaft, and a longer, but slower, angled drive shaft to deliver torque to the accessory gearbox. The intermediate gearbox accommodates the change in shaft angle between the radial drive shaft and accessory gearbox by the utilisation of a pair of spiral bevel gears. With modern engines, the intermediate gearbox is a line-replaceable module with easy access.

The angled drive shaft is housed within an oil-tight shroud tube, which in turn is protected from the bypass airflow by a splitter fairing. This arrangement has a similar function to the struts in the structures. As with the struts, it is important to minimise the performance losses associated with the splitter fairing, and a small diameter angled drive shaft helps to keep the splitter fairing as unobstructive to the fan airflow as possible.

Accessory gearbox

The accessory gearbox contains the drives for the accessories and the drive from the starter; it also provides a mounting face for each accessory unit.

The gearbox is crescent-shaped so that it wraps around the fan case. Wrapping the gearbox enables the nacelle to present a low frontal area, permitting improved streamlining of the surrounding engine cowl. The streamlining reduces drag when in flight.

Locating the gearbox on the underside of the engine allows the ground crew to gain access for maintenance. For the same reason, in helicopter installations, the gearbox is usually located on the top of the engine. The requirement to separate electrical units from fluid-filled units minimises the risk of fire. Electrical units are positioned on the 'dry' side of the gearbox, and fluid-filled units on the 'wet' side, separated by the starter, or input, gear shaft. The dry side generally mounts the generator (aircraft power) and permanent magnetic alternator (engine EEC power), and the wet side the hydraulic pump, fuel pump, and oil pump.

If any accessory unit fails and is prevented from rotating, it could cause further failure in the gearbox by shearing the teeth of the gear train. To prevent such secondary failure, the accessory drive shafts incorporate a weak section known as a 'shear-neck', which is designed to fail and so protect the other

drives. This feature is not included for primary engine accessory units, such as the oil pumps, because these units are vital to the running of the engine and their failure would necessitate immediate shutdown of the engine.

As the starter often provides the highest torque that the drive system encounters, it is typically the basis of the design. The starter is usually positioned to give the shortest driveline to the engine core. This avoids having to strengthen the entire gear train, which would increase the gearbox weight.

The gearbox provides two additional functions, those of the breather and the rotator. High-pressure air leaking through all the labyrinth seals in the bearing chambers must be exhausted by a device that retains the entrained oil. This is the function of the breather. The gearbox drives the breather at high speed. Oil-laden air flows into the breather, and the high rotational speed of the breather centrifuges the heavier oil from the air and returns the oil to the lubrication system.

Rotator provision is made for hand turning the engine during maintenance. This enables inspections to be performed on the rotating components of the engine by slowly turning

The bearing locations and gearbox arrangement on a three-shaft engine

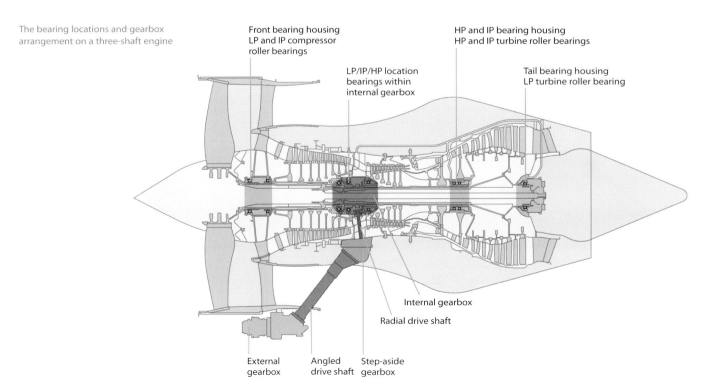

Front bearing housing
LP and IP compressor
roller bearings

HP and IP bearing housing
HP and IP turbine roller bearings

LP/IP/HP location
bearings within
internal gearbox

Tail bearing housing
LP turbine roller bearing

Internal gearbox

Radial drive shaft

External
gearbox

Angled
drive shaft

Step-aside
gearbox

The components of
an accessory gearbox

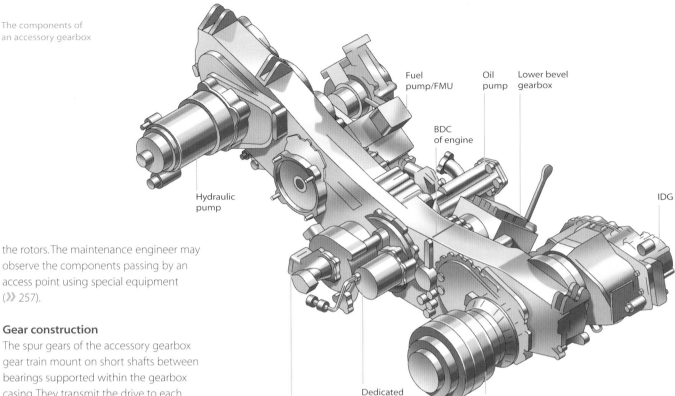

Fuel
pump/FMU

Oil
pump

Lower bevel
gearbox

BDC
of engine

IDG

Hydraulic
pump

Breather

Dedicated
alternator

Pneumatic starter

the rotors. The maintenance engineer may
observe the components passing by an
access point using special equipment
(» 257).

Gear construction

The spur gears of the accessory gearbox
gear train mount on short shafts between
bearings supported within the gearbox
casing. They transmit the drive to each
accessory unit, which may be as high as
15,000rpm for the accessory units and up
to 20,000rpm for the centrifugal breather.
Inside the gearbox casing, a line of parallel
gear shafts generally follows the curve of the
gearbox housing. The diameter of the gears
determines the spacing of the accessories.
An idler gear between adjacent accessory
gears provides additional space and
maintains the direction of rotation of the
drive shafts, generally clockwise.

Gearwheels from an accessory gearbox

Spur gears transmit power between parallel
axis shafts while spiral bevel gears transmit
power between shafts with intersecting axes.
The majority of gears within a gear train
are of the straight spur gear type; those
with the widest face carry the greatest loads.
For smoother running, helical gears are
used to improve the contact ratio, but the
resultant end thrust caused by this gear
tooth pattern must be catered for within
the mounting of the gear.

Gearbox sealing

Accessory gearboxes are provided with
lubrication from the engine oil system.
The accessory gearbox oil system is isolated
from any fluids present in the accessories,
such as hydraulic fluid or fuel, to prevent
cross-contamination. Sealing of the accessory
drive shafts is typically accomplished using
air-pressurised labyrinth sealing systems.
Within the accessory gearbox, at the
accessory mount pads, two sets of labyrinth
seal fins are statically mounted to the
gearbox housing in close proximity
to the rotating accessory drive shafts.
High-pressure air, fed centrally between

the sets of fins, prevents oil from escaping
from the gearbox. In the event of an
accessory failure, the air-blown seal prevents
contamination of the engine oil within the
accessory gearbox.

Gearbox materials

The gears are generally manufactured from
a forged stock of special gear steel, and
are carburise case-hardened for strength,
toughness, and wear resistance. After the
metallurgical processing, the gear teeth are
accurately ground for smooth gear meshing.

Strategically-placed oil nozzles provide
lubrication of gears during engine running.
Due to a momentary absence of oil flow at
the very beginning of engine start-up, it is
common to use a small amount of oil caught
on engine shutdown to lubricate the dry
gears. Another approach to providing start-up
lubrication is silver-plating the teeth. The silver
provides a soft, malleable, surface layer, which
acts as a dry lubricant. Silver-plating only one
of the gears aids the bedding in process by
allowing the uncoated 'harder' gear to polish
the silver coating on the mating gear.

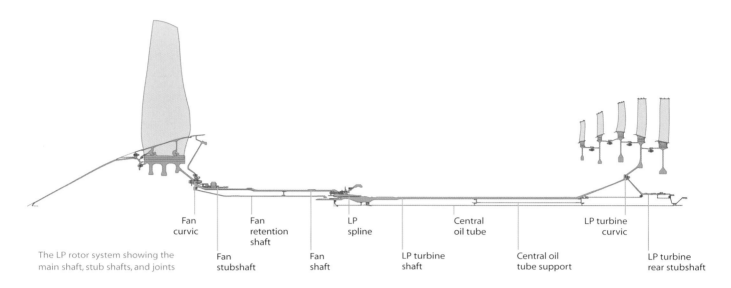

The LP rotor system showing the main shaft, stub shafts, and joints

Fan curvic | Fan retention shaft | LP spline | Central oil tube | LP turbine curvic

Fan stubshaft | Fan shaft | LP turbine shaft | Central oil tube support | LP turbine rear stubshaft

Shafts

Engine internal shafts are major parts of engine rotor systems. Their prime purpose is to transmit driving power from the turbine to the compressor end of a rotor. Within a three-shaft engine, the outermost rotor system is the HP rotor. Within the HP rotor, there are two more concentric shafts transmitting power – the IP and the LP.

The outermost shaft, known as the HP compressor drive cone, is large in diameter and short in length making the rotor system very stiff. Due to this stiffness, the HP rotor can be supported on two bearings: a ball bearing at the front taking the thrust, and a roller bearing at the rear. The next innermost shaft

Fan stub shaft assembly from a BR710

is the IP drive shaft connecting the IP compressor to the IP turbine. This shaft is longer and slimmer than the HP shaft, and renders the IP rotor too flexible to be supported only at the ends. Therefore it is supported in three places: in front of and aft of the IP compressor, and near the IP turbine. Finally, the LP shaft is innermost, connecting the LP turbine to the fan. This shaft is even longer and slimmer than the IP shaft, and the rotor is more flexible. Therefore, like the IP, the LP rotor needs three support bearings. They are located aft of the fan, aft of the IP compressor, and aft of the LP turbine.

The fan produces most of the engine thrust and absorbs most of the power. In addition, it turns at a slower speed than the IP and HP rotors. Being at the smallest diameter, it carries the highest stresses of any shaft in the engine.

Design considerations

Shafts are designed not to fail, but are also designed so that, in circumstances where it is clear the engine will not continue to operate, they will fail predictably and preserve the integrity of the airframe. For example, the LP turbine shaft is designed to deform in a predictable manner if a fan-blade-off event occurs. This enables the engine to shut down safely.

Shafts are also an important part of the air and oil systems, and allow the distribution of oil and air for lubricating and cooling. Because of the number of concentric shafts in a three-shaft engine, space at the centre

is limited. Therefore, air system holes (» 170) and the clearances between disc bores and shafts are especially critical in three-shaft engines.

Shaft materials

Shaft materials, especially the LP shaft, must strike a balance between high torque-carrying capability and high-temperature capability. High-strength steel alloys are often the choice, but these steels are not corrosion-resistant. Steel of this kind could corrode in service, particularly in the case of military transports that have intermittent use or are exposed to salt spray. To combat this corrosion, the steel is surface-blasted and coated with an aluminium epoxy paint.

Increasing bypass ratios bring the requirement for more torque through smaller bores, meaning that fatigue strength needs to be continually improved. Nickel chromium alloys can be used for turbine shafts – they are very expensive but have high fatigue and creep strengths, while also being corrosion-resistant. An alternative is to use steel alloys based on pure electrolytic iron with a very low sulphur and phosphorus content giving very high fatigue strength; however, their hardness brings additional machining challenges.

Shaft jointing

Shaft joints may have to carry a combination of torque, axial load, and bending moment. The three basic types of joint in use are bolted joints, splines, and curvic couplings.

A curvic coupling transmits torque from the turbines to the compression system

Bolted joints

These are the lowest cost and therefore most commonly used variety. Taper bolts are used to transmit torque through the joint, but they will not take significant axial load so a mixture of plain bolts and taper bolts may be required. Splined and curvic joints are preferred due to their higher torque-carrying capability, but are more costly than tapered bolts.

Splines

Splined joints are appropriate where radial space is constrained but axial space is available. The spline teeth are of involute form like a gear, but are stubby to withstand very high torque. The sides of the external splines are convex and the internal splines concave, making them self-aligning. Helical splines can carry torque, axial load, and bending moments, and remain self-aligning for all conditions.

Curvic couplings

Curvic couplings consist of interlocking rings of teeth between two adjacent discs, and are used where radial space is available but axial space is limited. The tooth flanks have a circular arc form making them self-centring. The interlocking teeth transmit torque from one disc to the next, but this action produces a separating force that tends to push the discs apart. To counter this bolts are needed to keep the mating curvics clamped together.

Bearings

Bearings provide a means of accurately locating the rotors while transmitting high forces with very little rotational resistance. Jet engines tend to use rolling element bearings, but occasional application of plain bearings can be found.

There are two types of bearing used in a gas turbine: ball bearings and roller bearings. Ball bearings use balls as the rolling elements, which, because of their shape, can withstand both radial and axial forces. This makes ball bearings suitable for transmitting thrust. Roller bearings use cylinders as the rolling elements. The rollers can transmit radial load across their diameters, but allow the shaft to slide lengthwise. Using a single ball bearing for thrust and one or more roller bearings to support a rotor allows positioning at the thrust bearing, but freedom for growth at the roller bearings.

Bearings can be used between rotating and fixed structures, or can be used between two rotating components. For example, the LP shaft thrust bearing on three-shaft engines is mounted between the LP and IP rotors. All rotating shafts in the engine, including the drive shafts from the internal gearbox to the accessory gearbox and the gear shafts within the accessory gearbox, are mounted on rolling element bearings.

All rolling element bearings consist of an inner and outer race, a cage, and the rolling elements themselves. One or both of the races have a raceway formed within it to guide the rolling elements.

The cage is used to maintain spacing of the rolling elements, which are trapped inside pockets. The cage has a clearance with respect to both the inner and outer races, but is primarily located by one or the other, depending upon the requirements of the bearing application. To ensure that the cage runs concentrically, the clearance between the locating lands on the race and the cage is small and well lubricated so that it operates without appreciable wear. The cage may also have features to assist in catching and directing lubrication to the rolling elements.

Ball bearings

Ball bearings provide axial location for a rotating shaft, but will usually carry a substantial radial load. A rotating shaft is supported by at least two bearings: normally, one is a ball bearing and the other, a roller bearing.

Main shaft location bearings are situated in the internal gearbox on three-shaft engines and on many two-shaft engines. Putting these highly-loaded bearings in a relatively cool part of the engine greatly simplifies design of the

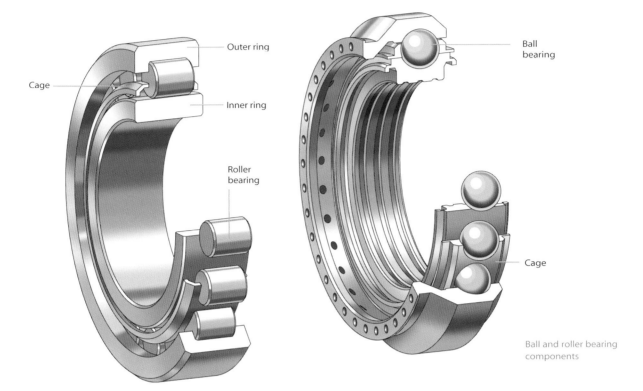

Outer ring

Cage

Inner ring

Roller bearing

Ball bearing

Cage

Ball and roller bearing components

load paths through the engine structures. Accurate axial location provided by the ball bearings is essential for close control of compressor tip clearances.

Deep groove
Deep-groove ball bearings have single-piece inner and outer rings. The cage is made, therefore, from two pieces to allow the bearing to be assembled. The inner and outer track forms are both derived from a single radius, and so the balls can only make single-point contact with each race. They are often used for applications with moderate radial loads and light axial loads.

Two-piece raceway type
This bearing commonly has a single-piece outer race and two-piece inner race, although it is possible to have a two-piece outer and single-piece inner. Splitting one of the races allows the bearing to be assembled and to have a single-piece cage. The raceway in each race is formed from two radii (one for each half of the raceway) struck from different centres so that the form of the track is a gothic arch. Since one of the races must be split, the thrust load must be maintained at a high level during operation to prevent the balls from contacting the split. Therefore,

View of an assembled ball bearing, also called a thrust or location bearing

these bearings are used in more applications that require high thrust-carrying capacity.

The gothic arch form allows oil to be fed into the centre of the inner track without the risk of damage that might result from the balls running over the edges of the oil feed holes. Supplying oil to the centre of the inner race gives good lubrication at the ball contacts. This configuration is the most commonly used for main shaft location bearings, as used on the Trent LP, IP, and HP main shafts.

Roller bearings
Roller bearings are used in all main shaft and auxiliary drive shaft applications to

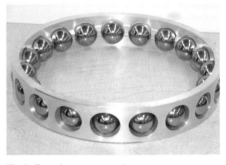

The balls and cage removed from the inner and outer race

support pure radial load, and allow for axial shaft elongation due to temperature changes with no additional load effect on the bearing. They are usually located at the ends of the turbine and compressor shafts and are often mounted in a housing, but separated from it by a layer of pressurised oil known as a squeeze film damper.

In many cases, instead of having a separate inner race for roller bearings, the inner race is an integral part of the shaft or stub shaft. This reduces complexity, weight, and build-up of concentricity tolerances. Overall, this is cost effective, but the cost of replacement or repair is likely to be higher than for separate inner races.

An LP fan roller bearing from a three-shaft engine

Bearing internal clearance

Bearing diametral clearance is the total free movement between the inner and outer races in the radial direction. For ball bearings, there must be some positive diametral clearance under all operating conditions. Roller bearings and ball bearings that are mainly radially-loaded benefit from low diametral clearance. This maximises the number of loaded elements and reduces rolling element-to-race stress levels. For roller bearings, a low diametral clearance also helps to reduce the risk of roller skidding.

Bearing squeeze films

In some engines, a squeeze film is used to minimise the dynamic loads transmitted from the rotating assemblies to the bearing housings. Bearing squeeze films are small, oil-filled clearances between the outer race of the bearing and its housing. The oil film dampens the radial motion of the rotating assembly and the dynamic loads transmitted to the bearing housing, thereby reducing the vibration level of the engine and the possibility of damage by fatigue. Oil is retained in the film space by either a close axial clearance in the bearing housing or by a piston ring seal at each end of the film. The squeeze film also alleviates some of the effects of engine carcass deflections on the shaft, caused by manoeuvre loads or asymmetric thermal expansion. When a squeeze film is applied to a shaft thrust bearing, flexible bars are used to

attach the bearing outer race to the static structure to carry the axial load while still allowing radial movement and shaft centring.

Bearing materials

Bearings are currently manufactured from steels that may be either case-hardened or through-hardened to suit the application. Rolling element bearings operate with high local stress levels at the contacts between the rolling elements and the races. This means that the material used must have a very high resistance to rolling contact fatigue. Other requirements of the material are a high level of hardness at the surface, high temperature and wear resistance, and often a tough core.

The effects of rotation and installation fits can further increase these stress levels. Surface-hardened materials have an additional attribute: a surface that is usually in compression. This is beneficial to a surface in tension and tends to cancel out the effects of rotation and fit. Corrosion resistance and damage tolerance may be other important attributes in some applications.

Most bearings employ high quality steels for the cage material. However, lower duty bearings may use phosphor bronze or brass cages. Silver plating and phosphate coating enhance friction, lubrication, and wear properties on steel cages.

Bearing developments

The demands for future gas turbine bearings will be longer life, higher speeds, higher load capacity, smaller diameters, and (for aero engines) less weight. Steel processing continues to improve and is delivering cleaner, inclusion-free materials, leading to higher fatigue resistance.

Current technology goes some way to meeting these needs. However, alternative materials such as ceramics, polymers, and composites will play a future role in aerospace bearing technology, particularly in high-speed applications. They offer high strength for low weight and work well in high temperatures and poor lubrication conditions. Specialist surface treatments are also being developed that will enhance bearing performance.

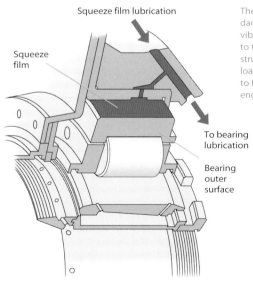

Squeeze film lubrication

Squeeze film

To bearing lubrication

Bearing outer surface

The squeeze film damper reduces vibration transmission to the engine structure, and reduces loads on the shaft due to flexing of the engine casings

If transmissions provide the skeleton of the engine,
fluid systems are its life-blood.

fluid systems

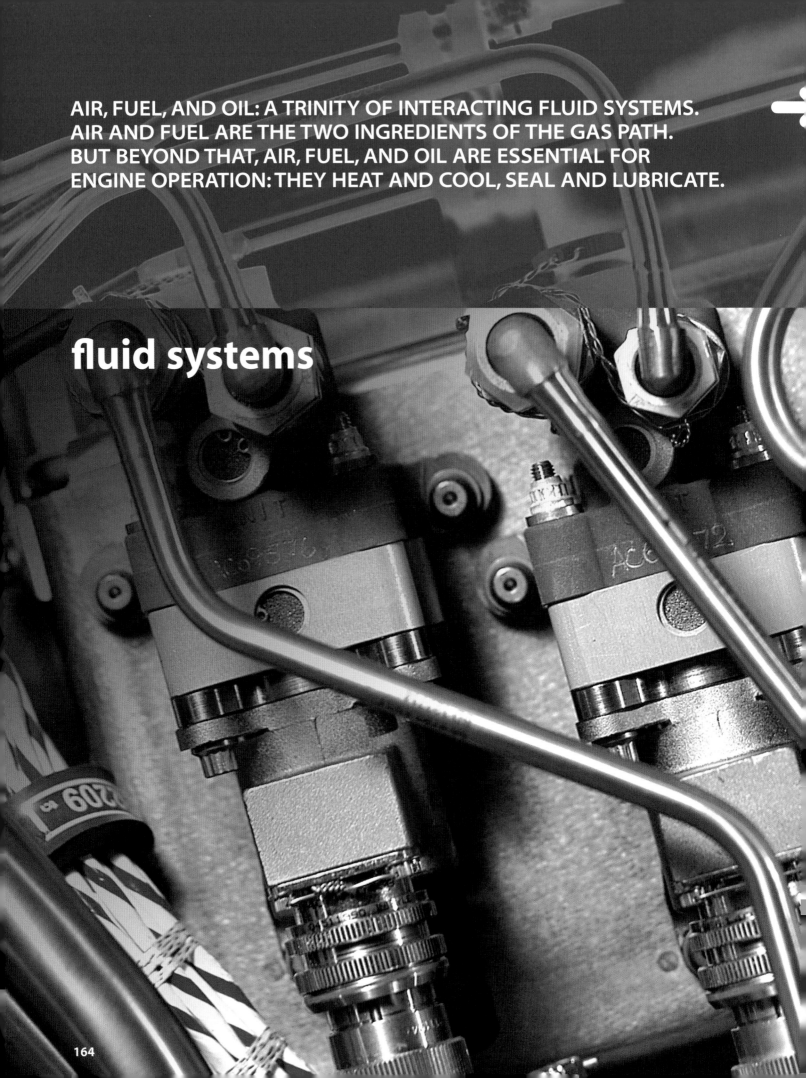

AIR, FUEL, AND OIL: A TRINITY OF INTERACTING FLUID SYSTEMS.
AIR AND FUEL ARE THE TWO INGREDIENTS OF THE GAS PATH.
BUT BEYOND THAT, AIR, FUEL, AND OIL ARE ESSENTIAL FOR
ENGINE OPERATION: THEY HEAT AND COOL, SEAL AND LUBRICATE.

fluid systems

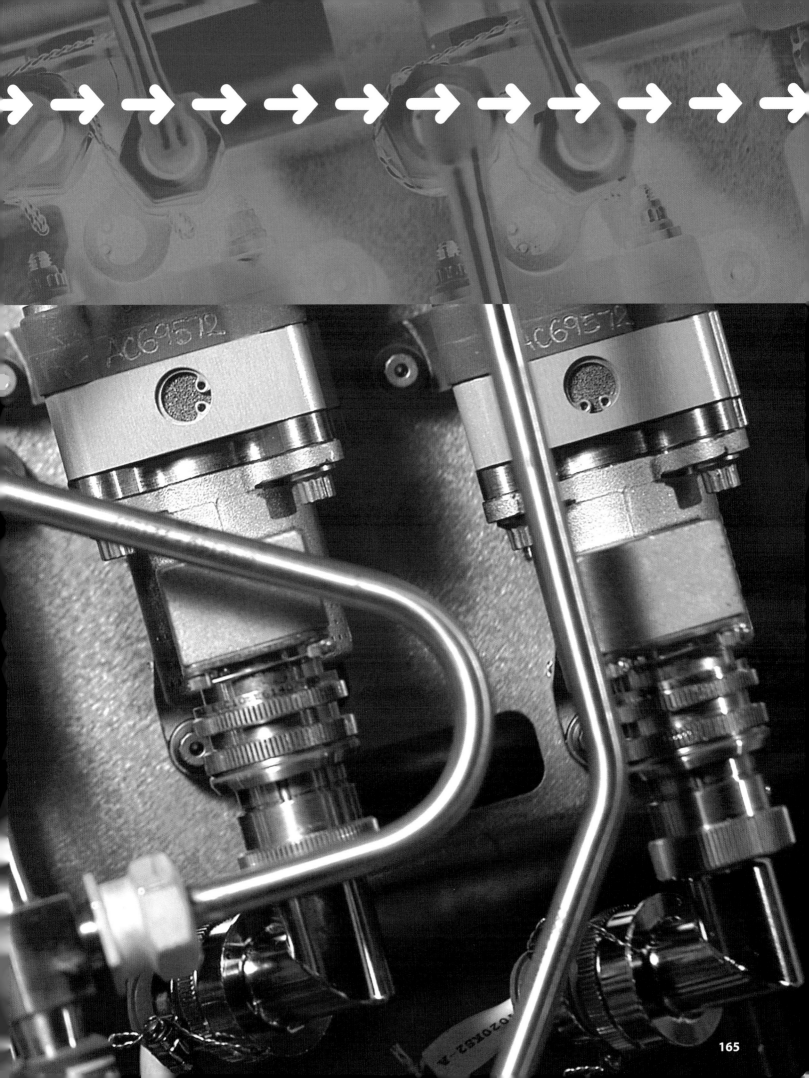

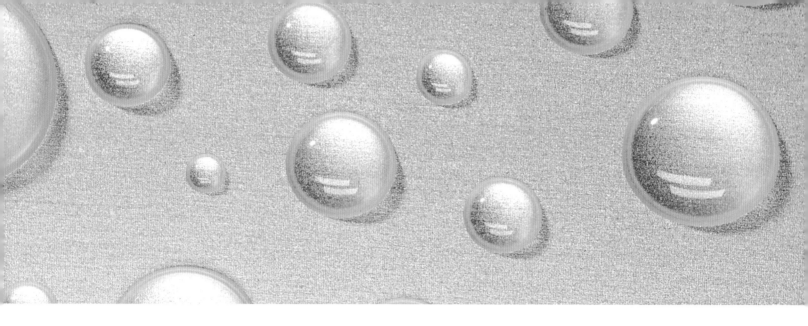

The functions of the internal air system include

❭ providing a cooling flow to engine components

❭ sealing bearing chambers and flowpaths

❭ controlling bearing axial loads.

Up to 20 per cent of the engine core flow may be used for these functions. This can be equivalent to five per cent of the energy available in the fuel consumed – a very significant cost for the engine operators.

The fuel system is designed to provide

❭ an uninterrupted supply of fuel to the combustor and reheat systems, as demanded by the engine thrust management and control systems

❭ a source of hydraulic power to actuate control system variables, as demanded by the thrust management and control systems

❭ a heat sink for the oil systems and electrical generating systems on the engine.

The oil system of an aero gas turbine provides

❭ lubrication

❭ cooling

❭ corrosion protection.

The three fluid systems interact with each other at various points in their cycle through the gas turbine.

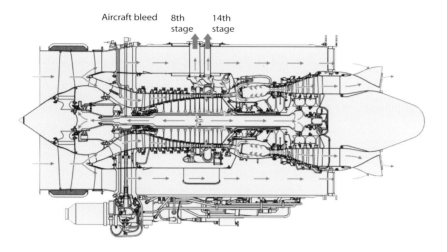

Aircraft bleed 8th stage 14th stage

The general arrangement and flowpaths of the internal air system of the AE 3007 two-shaft engine

Air systems

Air – the working fluid in a gas turbine engine – is compressed, heated, and expanded to produce power. Some of the compressed air does not contribute directly to the production of thrust or shaft power. Instead, it is used for functions vital to the safe and efficient operation of the engine: cooling, sealing, and controlling bearing loads. These secondary airflows, and the collection of hardware features that direct the airflow paths, define the engine internal air system.

Cooling

Several areas of the engine require cooling to maintain safe operation – most of all, the combustor and turbines as they experience the highest heat loads. The combustor is cooled by the gas path, not the internal air system.

Air extracted from the compressor discharge cools the HP and, where necessary, the IP turbine components. The cooling air can be over 700°C – enough in itself to melt most aluminium alloys – while the mainstream gas temperature in the parts of the turbine can be over 1,600°C, necessitating the use of high-strength, high-temperature superalloys in these areas.

Cooling turbine blades and nozzle guide vanes

The gas turbine engine thermal efficiency increases with the turbine entry temperature, TET – a fact of the thermodynamic cycle. The higher heat load from running the engine at higher TETs means that cooled aerofoils are used, accounting for a large

The turbine cooling circuit in the AE 3007

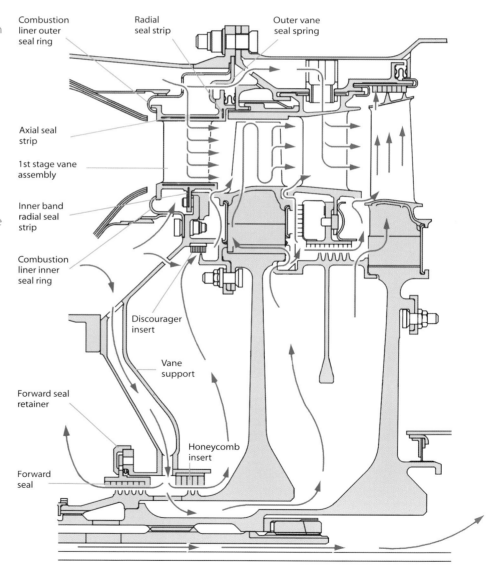

Combustion liner outer seal ring — Radial seal strip — Outer vane seal spring — Axial seal strip — 1st stage vane assembly — Inner band radial seal strip — Combustion liner inner seal ring — Discourager insert — Vane support — Forward seal retainer — Honeycomb insert — Forward seal

portion of the total cooling flow usage in the engine (» 138). The performance gains with increasing TET then become limited due to the negative performance impact of increased cooling flows.

Modern cooled aerofoils incorporate a variety of schemes in an effort to maximise the cooling effectiveness – with cost increasing in proportion to the complexity of the cooling scheme. The evolution of cooled aerofoils currently favours multiple feeds, multiple passes, and extensive film cooling. Advanced materials can simulate a porous media and allow a TET getting closer and closer to that of stoichiometric combustion (» 116, 126).

Cooling turbine discs and shafts

Discs and shafts are typically classified as critical parts and their integrity must be maintained under all conditions; this limits a disc's maximum operating temperature.

Discs and shafts are heated by conduction of heat from the mainstream gas path through the blade-disc contact area. Cooling air flows axially across the bore of the disc and radially over the disc faces. The heat capacity of the disc combined with the heat transfer between the air and disc surfaces create temperature gradients through the disc during the acceleration and deceleration portions of the engine cycle. The resulting thermal stress from the alternating, non-uniform expansion and contraction of the disc material is a component of the total stress that determines the disc's cyclic life. Optimising the cooling airflow to the discs increases the life of the disc.

Cooling turbine casings

Air is supplied to turbine casings for three reasons:

> as part of the delivery path for nozzle guide vane cooling

> to provide cooling to maintain casing material strength

> to control the thermal growth of the casing, thereby controlling the clearance between the blade tip and casing during transient operation.

For unshrouded blades, blade tracks fixed to the casing help maintain close blade tip clearances for maximum turbine efficiency. These tracks are cooled with techniques similar to those used for blade and vane internal cooling. The air system is designed to prevent the ingestion of mainstream gas into the blade track cavities. Air flowing through casings also contributes to outer surface (skin) temperatures, which must be kept below the ignition temperature of the fuel.

Cooling accessories

Some engine-driven accessories (for example, the electrical generator) generate a significant amount of heat that must be dissipated to keep the unit at an acceptable running temperature. A lower stage compressor off-take may be used to supply cooling air directly to the unit; another method is to cool the unit with atmospheric air. This is achieved by allowing compressor delivery air to pass through nozzles in the cooling air outlet duct of the accessory. The air velocity through the nozzles creates a low pressure area, which forms an ejector, so inducing a flow of atmospheric air through the intake louvres.

Sealing

Sealing aims to minimise the performance penalties from air leaking overboard, across engine modules, and across turbine stages. The air system includes seals between rotating and static parts, co-rotating and contra-rotating parts, and static parts.

The internal air system must provide effective sealing in order to direct cooling air to the target locations at the designed flow levels. Excessive leakage may require changes in the air system architecture for it to perform correctly – for example, using a higher stage compressor bleed.

Preventing oil leakage is an important sealing function. Oil leakage outside the bearing chamber may result in an engine fire. A leak into the mainstream gas path may cause aircraft cabin odour, or visible smoke – an especially alarming event. Air is used to buffer seals around bearing chambers to prevent oil leakage, but too much airflow is a performance penalty and increases the heat load to the oil in the chamber.

Another key sealing function is minimising mainstream gas ingestion into the turbine rim cavities. The air system must provide enough cooling flow either to purge the rim cavities, so preventing ingestion, or at least to dilute the hot gas within the rim cavities enough to achieve an acceptable temperature level.

Control of bearing loads

The flow of the mainstream gas exerts an axial force that acts in the forward direction on the compressor, and in the aft direction on the turbine. The shaft connecting the compressor and turbine will experience a net axial load that is the sum of the compressor and turbine gas loads, and the loads produced by the internal air system acting on the discs and shafts. The position of sealing elements around the compressor

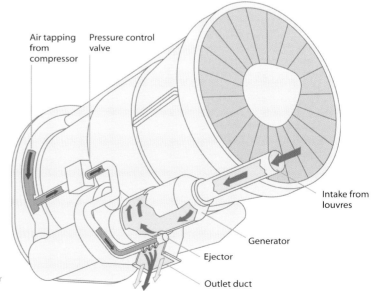

Air tapping from compressor

Pressure control valve

Intake from louvres

Generator cooling uses compressor air and an ejector to induce intake air through the generator

Generator

Ejector

Outlet duct

and turbine determine the net internal air loads and is the primary tool available to the air system designer for controlling bearing loads. Another important factor in the resultant load is the reaction of the HP turbine. The reaction determines the gas path static pressure between the first stage nozzle guide vane and blade. This pressure can act over a large area of the disc and change net axial loads significantly. The bearing loads must be controlled to reduce the risk of overloading or unloading a thrust bearing. An unloaded bearing is more likely during engine operation, and the rolling elements can skid when unloaded causing significant heat generation leading to bearing failure.

Customer and external bleeds

Substantial quantities of air are bled from one or more stages of the compressor for aircraft services including cabin pressurisation, cabin heating, and airframe anti-icing. The aircraft control system determines the demand for the bleed air and will take the lower stage bleed as long as the delivery pressure is adequate, switching to the higher stage bleed for low power points in the flight envelope. In this way, the performance penalty on the engine is minimised.

Customer bleeds, though taken from the compressor outer casing and routed outside the engine, affect the internal flow system by changing the compressor operating point. If the customer bleed and cooling bleed have a common off-take stage, the pressure available to the internal system changes with the customer bleed demand. There are similar issues with other external bleeds, such as engine anti-icing or accessory cooling. Bleeds for starting and handling bleeds for compressor surge avoidance are discussed in the context of engine operability.

Air system elements
Fixed areas – holes, slots, and ducts

For air to flow from one point in the engine to another, a flow area must be created. The simplest example of this is a circular hole drilled in a stationary wall separating two regions at different pressures. Even for this most basic case, the amount of air that can pass through that hole depends on many factors.

Effects arising from the finite viscosity and compressibility of air determine the actual mass flow rate. In measurable terms, the flow rate depends on the geometry of the hole (thickness, shape, and profile), the ratio of the upstream and downstream pressures, and the upstream air temperature. Air velocity is also a factor: it may not be parallel to the hole axis, or the hole may be in a rotating component, which is rotating at a different speed from the whirl velocity of the incoming air.

The discharge coefficient (C_d, the ratio of the actual to the theoretical mass flow rate) is often used to describe the flow through a fixed area.

Rotating seals

The rotating seal is placed between two parts, one or both of which are rotating. If both components are rotating, there can be a difference in rotational speed and direction. A close clearance between a shaft and a bore can be considered a seal, but modern rotating seals include special features that help minimise leakage by creating pressure losses and, thus, a resistance to airflow. Seals must also cope with the relative axial and radial movement between the rotating components during the flight cycle.

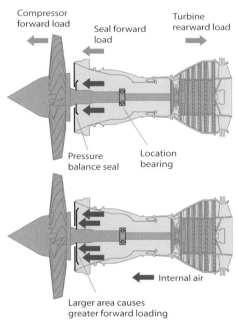

Axial bearing load control through setting the pressure acting on disc faces

Control of bearing axial load

Compressor forward load

Seal forward load

Turbine rearward load

Pressure balance seal

Location bearing

Internal air

Larger area causes greater forward loading

Labyrinth seals

Labyrinth seals are widely used in gas turbine engines for all sealing functions of the air system. The basic labyrinth seal creates a resistance to airflow by forcing the air to traverse through a series of fins. The fins run close to the seal's outer lining, and pressure losses are generated by the acceleration and expansion of the air as it passes between each fin tip and the lining. Enhancements such as inclining the fin into the flow, and radially stepping up or down successive fins, will improve seal performance, usually at a greater cost and space claimed by the seal.

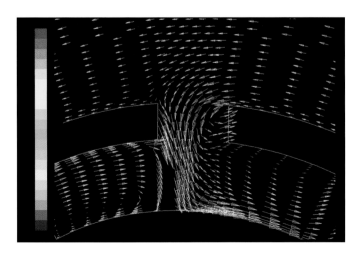

Velocity vectors of swirling air passing through a hole in a rotating shaft – the larger the difference in speeds, the smaller the discharge coefficient, C_d

Fluid and abradable lined labyrinth seal

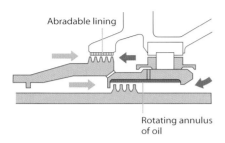

Abradable lining

Rotating annulus of oil

Continuous groove interstage (labyrinth) air seal

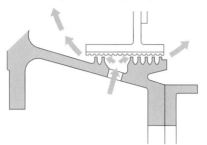

Intershaft hydraulic seal

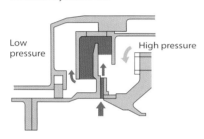

Low pressure

High pressure

Ring type oil seal

Brush seal

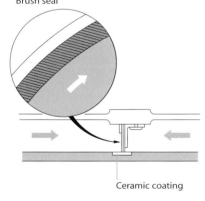

Ceramic coating

Typical carbon seal

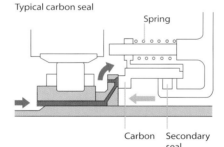

Spring

Carbon

Secondary seal

 Sealing air

Oil

Rotating assemblies

To cope with the relative radial movements, the labyrinth seal may be designed in such a way that the fin tips never touch the outer lining, or a soft, abradable material or honeycomb structure may be included on the outer lining that is designed to tolerate rub. During the initial running of the engine, the fin tips rub into the material and cut grooves to the deepest radial extent seen transiently. Thereafter, the fins tend not rub further but excessive shaft movements caused by aircraft manoeuvres or hard landings will cause the fin to rub occasionally. Generally, the running clearance in a labyrinth seal gradually increases throughout the engine service life.

In co-rotating shaft applications, the abradable lining may be replaced by a rotating annulus of oil for bearing chamber seals. As the shafts deflect, the fin tips enter the oil, and seal performance is maintained without the heat generation produced by rubbing a metallic lining.

Brush and leaf seals

Brush seals consist of a static ring of densely packed, fine wire bristles (usually metallic) that are angled in the direction of rotation of the rotating component. The bristles are in continuous contact with the rotating member, rubbing against a hard ceramic coating. Very low leakage is possible with this type of seal. The compliant bristles take up any assembly misalignments as well as relative radial movements during engine operation. The brush seal is not generally used to seal bearing chambers since broken bristles could contaminate the oil supplied to the bearings, and lead to premature failures.

Leaf seals work on the same principle as brush seals but are made from foil rather than wire. This seal has higher axial rigidity and is less susceptible to backing plate wear.

Carbon seals

Carbon seals are generally used for sealing oil within bearing chambers and gearboxes. They rely on a positive pressure differential to load the carbon elements adequately – although this is frequently supplemented by various compression springs. One or more carbon elements make up a static but conforming ring positioned between a static component and a rotating surface.

171

These seals normally require oil cooling as any contact between the carbon elements and the rotating surface generates considerable heat.

Air-riding carbon seals are designed to operate with minimal or no oil lubrication or cooling. Small scallops machined into the contacting surface allow the carbon elements to lift off and run on a small cushion of air.

Ring seals

Ring seals are used to seal bearing chambers by forming a close clearance between the static ring and the rotating shaft. The ring is loosely contained in its housing so that the ring can move when the shaft deflects and contacts the ring. Binding of the ring in the housing can occur in high-temperature environments due to fouling of the oil.

Hydraulic seals

This seal is used to seal bearing chambers in co-rotating shaft engines. This is an essentially zero air leakage seal, unlike other seals used for bearing chambers.

A rotating annulus of oil is created in the outer member by centrifugal forces. A fin on the inner member rotates, immersed in the oil, forming the seal. A difference in the air pressure outside and inside the chamber is compensated by a difference in oil level either side of the fin. The oil rotation speed is very close to the outer ring speed, and any speed differential creates oil shearing and heat. To control the heat, it is usual to have oil flowing through the seal from the high-pressure side.

Static seals

Static seals are used between structures that cannot otherwise feature positive sealing (such as a clamped joint or interference fit) because of assembly requirements or small relative movements due to thermal expansion. Examples include the interfaces between turbine casings, vane segments, blade platforms, and blade shrouds. The term 'static' refers to the relative movement of the components' surfaces being sealed, so both could be rotating together.

Cavities

The many cavities formed by the turbine and compressor discs and static structures include major portions of the engine cross-section and important features of the internal air system. These cavities form conduits through which air is delivered. The air system requirements do not strongly influence the size, shape, and arrangement of these cavities, but the effects of the cavities on the air system performance must be considered; features are adjusted to take advantage of, or compensate for, these effects.

Flow through the cavities produces changes in air pressure, temperature, and whirl velocity. The extent to which these properties change depend on the level of flow rate, the disc speeds, whether the air is flowing radially inward or outward, and whether the cavity is formed by two rotating discs or a disc and static structure. While the net flow through these cavities is important for the overall balance of the air system, it is increasingly important also to understand the entire flow field within some cavities in order to quantify the heat load distribution on the cavity walls

Whirl velocity control

Changing the tangential (whirl) velocity of the air, either increasing it (preswirl), or decreasing it (deswirl) is a very important tool in the air system.

The primary air source for cooling the HP turbine blades is HP compressor discharge taken directly from the diffuser. This air has no whirl velocity, yet must be delivered to blades that are rotating at the HP shaft speed. If this air is supplied with no whirl velocity, the disc must do work on the air, heating it up, to get it rotating at the disc speed. Hotter cooling flow results in hotter blade temperatures or increased cooling flow requirement.

Preswirling the air avoids this temperature increase and can be worth as much as 50°C in cooling supply temperature. This is a significant figure considering that creep life of some materials is halved by a 14°C increase in temperature (» 139, 144).

Preswirling is achieved by forcing the cooling flow through nozzles angled in the direction of rotation. More whirl can be achieved at the expense of pressure drop through the nozzle. Preswirl nozzles are either drilled holes or aerofoils. The interaction of other parts and cavities may make it more desirable to locate the nozzles at a higher radius.

Reducing the swirl of an airflow is sometimes done to extract energy from the air to improve the performance of the engine. Slots in turbine disc spacers, angled in the opposite direction to the rotation, are designed for this purpose.

Deswirling can recover pressure in compressor drum bleed flow. Some air is bled inward, between compressor discs, from an intermediate stage of the compressor. The air flows radially inward, and the pressure and temperature decrease as the whirl velocity increases due to the conservation of momentum principle.

Fixing tubes between the compressor discs at the air off-take location so that the whirl velocity of the air is forced to the speed of the tube, and therefore disc, rather than increasing freely between the discs, allows pressure to be maintained. The negative impact of deswirling is the added weight and cost of the deswirl tubes and associated disc features.

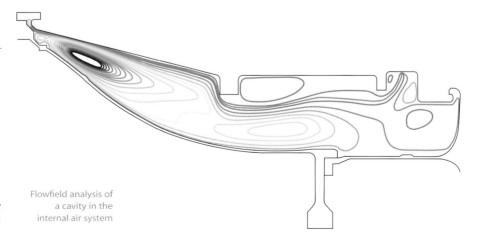

Flowfield analysis of a cavity in the internal air system

Air system design operating envelope

Throughout the service lifetime of the engine, the engine internal air system must perform its functions properly over the operational envelope defined by the customers and operators of the engine.

Engines on civil aircraft operate at altitudes ranging from sea level to 15,500m (51,000ft), at a range of power settings, and at varying aircraft speeds from static to Mach 0.92. The aircraft may require compressor air bleed anywhere within the flight envelope. Military aircraft operations expand the required altitude and speed ranges, and military engines generally run at higher power for longer periods, as a percentage of the entire flight.

Aircraft manoeuvres affect air system performance primarily by causing deterioration of the air seals. Hard landings and manoeuvre loads may cause shafts and static structures to deflect more than usual allowing labyrinth seal fins to rub against their linings.

Ambient temperatures range from −54°C to +54°C and the resulting changes in engine performance has an impact on the internal air system. More important, however, is the amount and type of debris in the air: sand, dirt, and soot when ingested into the engine may clog internal air passages, foul contacting seals, or block film cooling holes in the turbine nozzle guide vanes.

The design challenge

The design goal for the internal air system is to select, often from several possibilities, the most robust system architecture in the face of numerous challenges:

❭ satisfying customer requirements

❭ accommodating a large operational envelope

❭ tolerating failure modes

❭ reducing risk.

Customer requirements are foremost when designing the internal air system. Understandably, there is the desire to do more with less and great efforts are made to use the lowest stage compressor air and

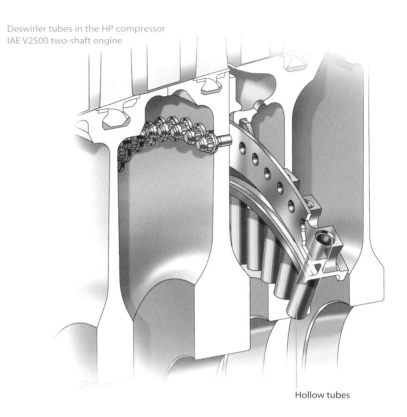

Deswirler tubes in the HP compressor IAE V2500 two-shaft engine

Hollow tubes

minimise leakages. Anticipating growth variants of the engine is important so that flexibility can be built into the air system design.

The operational envelope presents the air system with variable source and sink pressures and temperatures, shaft speeds, and seal clearances – all of which must be considered when selecting the type and location of the air system elements.

Failure modes add another dimension to the air system design challenge. It is required that no single point failure can cause a catastrophic engine event. It must be shown that failures of certain air system components can be tolerated or at least recognised before a safety issue arises.

Ensuring air system integrity

The design challenges described above are met, and the integrity of the design ensured, by several means including analysis of various component or engine types, engine and rig tests, referring to past experience and lessons learnt, and strict adherence to the design review processes that lead to formal certification (❭❭ 42 – 51).

BR710 vortex reducer assembly

Advanced network models are constructed and used throughout the life-cycle of the engine. These models simulate the entire internal air system at the critical points in the operational envelope. In addition to simulating design and off-design points, the models are used to simulate failures of certain air system elements.

A full internal air system pressure and temperature survey with actual engine data is required as part of an engine development. Analysis models are correlated to the measured data; often, only one such test is required.

Air system health monitoring

The primary on-wing health monitoring method for the air system monitors the normal engine performance measures: gas temperature and fuel economy. Secondary measured parameters can sometimes suggest degradation of air system elements. In certain engines, main oil pressure can be affected by seal wear and unusually high main oil pressure or a high rate of change in pressure can indicate an air seal problem.

Fuel system

Modern digital computer technology, in the form of a Full Authority Digital Engine Control (FADEC) system, has offered the opportunity to greatly reduce the complexity of hydro-mechanical and pneumatic engine systems while adding flexibility for the aircraft. The electronic engine controller (EEC) is the central control intelligence in a FADEC system (» 197) with EEC software replacing most of the hydro-mechanical and pneumatic elements of the fuel system.

A FADEC fuel system consists of a low-pressure (LP) circuit and a high-pressure (HP) circuit. Fuel is provided from the aircraft fuel tank to the engine LP fuel pump via the aircraft fuel system. The LP fuel pump provides the pressure to overcome the losses in the LP fuel system and supply pressurised fuel to the HP fuel pump.

The fuel oil heat exchanger (FOHE) provides oil cooling and fuel heating. The main LP filter protects the HP pump and the other downstream units from contaminants in the fuel. The HP pump provides sufficient fuel flow above the combustor pressure to satisfy the engine demand.

The fuel metering unit (FMU) controls the engine-consumed flow in response to the EEC demand. The FMU is also a servo pressure source for a remote actuator to operate the variable inlet guide vanes (VIGVs) of the HP compressor. The engine fuel flow transmitter generates an output signal proportional to the mass of fuel going through it. The HP filter provides the final protection for the fuel spray nozzles. The burner manifold distributes fuel to the fuel nozzles, which atomises the fuel for the combustion process.

Fuel system operation

A typical flight consists of a number of distinct phases to be considered when designing the fuel system.

On start up, the FMU metering valve and shut-off valve are opened, allowing the fuel flow delivered by the pump to pass to the fuel nozzles for ignition of fuel in the combustor without any metering of flow – open loop control.

Following start up, the fuel flow is metered and the system is controlled in a closed loop. The engine is run at an idle condition while the aircraft taxis, and the engine oil system is warmed to a desired temperature for acceleration to take-off.

When take-off power is demanded, the fuel system is capable of delivering maximum required fuel flow at maximum pressure. As the aircraft climbs following take-off, the required fuel flow and pressure reduces until the desired cruise altitude is reached.

At cruise altitude, the local ambient air temperature can result in fuel temperature dropping to around –35°C due to cooling over a long cruise. This fuel cooling requires detailed consideration in the fuel system design phase.

The outside air temperature at maximum altitude can be below –60°C; however, the fuel tank temperature does not approach the outside air temperature, rarely going below –35°C. This is due to heating of the wing structure caused by the aircraft airspeed. On hot days, the maximum fuel temperature can reach 55°C.

After the cruise phase, the engine power setting is reduced to allow the aircraft to descend and land.

Over the full flight, the fuel system is designed to ensure that a minimum desired fuel pressure is achieved, so that VSV actuators can be actuated to aid in the performance of the engine; this minimum requirement is typically not a concern at take-off where fuel pressure is high. However, at cruise and descent, the required fuel pressure may have a significant impact on heat generation – therefore raising fuel temperatures.

Due to the interconnectivity between the oil and fuel system by use of the FOHE, and the wide range of fuel flows experienced during

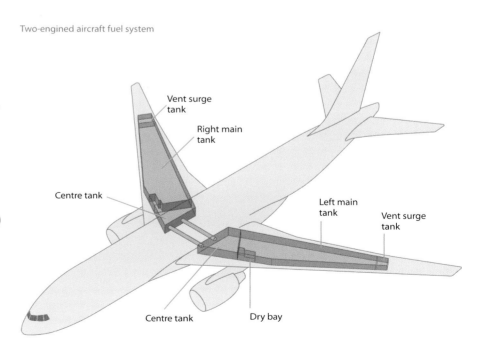

Two-engined aircraft fuel system

Vent surge tank

Right main tank

Centre tank

Left main tank

Vent surge tank

Centre tank

Dry bay

flight, the fuel temperatures can vary significantly between different phases of the flight. The resultant fuel and oil temperatures experienced are a key aspect in the design of the fuel system and its heat management.

Aircraft fuel system description

Fuel storage

Each main wing tank feeds its associated engine. If a centre tank is used, it typically feeds all engines and empties first; this assists aircraft aerodynamics and reduces risks in case of an emergency landing. Fuel shut-off valves for each main engine feed line are installed at the tank outlet to isolate the engine fuel supply. The tanks are vented to atmospheric pressure to permit equalisation of the tank pressure differential that is created due to changes in altitude or during pressure refuelling or defuelling.

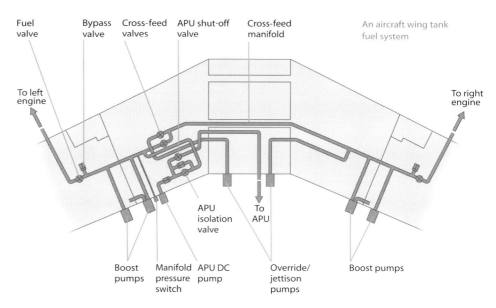

An aircraft wing tank fuel system

Fuel distribution

Each tank typically contains a pair of electrically-driven booster pumps that are either located in the inboard end of each main tank, or in a separate compartment, called the collector cell. This collector cell is always kept full of fuel via a fuel transfer system, preventing the pumps from becoming un-submerged during negative-g conditions.

The fuel transfer system enables the fuel to be transferred from any one tank to another in case of fuel asymmetry, or if one engine is to be supplied from its opposite tank group (known as 'cross-feed'). Each engine has a dedicated fuel feed system that is independent of any other but which can be interconnected.

The tank boost pumps are controlled in the cockpit. The aircraft-to-engine feed network is an important consideration due to the pressure losses caused by pipe losses and pipe height changes. With electric tank pumps, the engine inlet pressure is significantly above the minimum required inlet pressure ensuring the fuel is neither aerated nor contains free vapour.

Fuel system indication

The tank content is measured and fuel level switches are installed in the tank to provide fuel tank content indication to the flight deck.

Fuel tank temperature is measured so that a caution message can be provided to the flight deck if the tank temperature either decreases, as can be caused by long-range high altitude flights, or increases, possibly indicating heated fuel returning to the tank system. The boost pumps have pressure switches to provide status indication to the flight deck.

Aircraft and engine interaction

When the engine is installed on an aircraft, the two fuel systems function as one; their interactions are considered during both aircraft and engine fuel system design.

Suction operation

The LP fuel pump provides sufficient pressure rise to ensure that the HP pump can deliver the demanded fuel flow. In the event of the aircraft boost pumps being inoperative, the ability of the LP fuel pump to provide the necessary pressure rise is reduced if the fuel contains an excessive mix of air and fuel vapour. Two factors can cause such a mix of fuel, air, and vapour: first, the release of dissolved air in the fuel, and, second, fuel vaporisation caused by low fuel pressure as a result of aircraft fuel system pressure loss and low ambient air pressure at the aircraft altitude.

The LP pump is designed to ensure that the fuel flow required by the engine can be

provided over a suitable operating envelope. This envelope is determined during aircraft flight tests.

Negative-g conditions

Severe aircraft manoeuvres can lead to an interruption in the fuel flow to the engine fuel system, which in turn could cause a flame-out and loss of thrust.

Priming, re-priming, and relight

Several factors may result in an interruption to the fuel flow to the engine: suction operation and negative acceleration, as described above; also, ingestion of air by operation of the aircraft cross-feed, and interruption of fuel by accidental closure of the aircraft-to-engine fuel feed valve. Under all these circumstances, the engine may flame out. Following flame-out, the fuel system must be able to provide the engine with the required fuel flow to allow engine re-light, and normal engine operation to resume.

Pressure spikes

Fuel pressure spikes ('water hammer') are created whenever fuel flow is altered. When there is a large change in fuel flow occurring very rapidly, the fuel pressure spike magnitude can be very large (positive and negative). The pressure spike can be experienced in the aircraft and engines due to the inter-connectivity of the fuel systems.

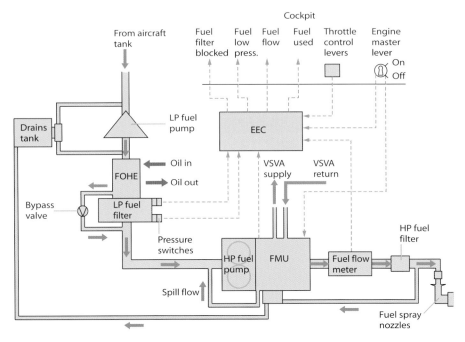

Schematic of a typical aircraft fuel system

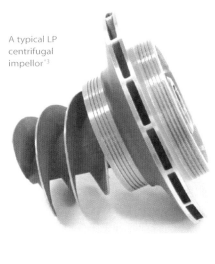

A typical LP centrifugal impellor[3]

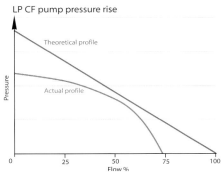

LP CF pump pressure rise

LP centrifugal pump pressure rise – actual and theoretical

Contamination

Since the aircraft fuel tanks provide the engine fuel, the engine can be presented with fuel-borne contaminant within the aircraft fuel tanks. This contaminant can be not only solid particles of dirt, dust, or debris, but also water or ice.

FADEC engine – fuel system description

Low-pressure pump

The purpose of the LP centrifugal pump is to maintain the fuel pressure at the inlet of the HP pump at a value high enough to prevent cavitation.

The LP pump produces a pressure rise based on the speed of rotation of the impeller. The pump design has to consider not only normal operation, but also the event of aircraft boost pumps failure. To cater for this situation, an inducer enhances the pump performance under operation with fuel containing air and vapour mix.

The LP pump is often also a pressure source to operate a small ejector pump, which is part of the fuel drains system. Fuel drained from the fuel manifold during engine shutdown is stored in a drains tank, and transferred back to the engine LP pump inlet by the ejector pump following the next engine start.

Fuel oil heat exchanger

The fuel oil heat exchanger (FOHE) extracts heat from the engine oil providing oil cooling and fuel heating. The unit is typically a shell and tube type flow heat exchanger. Fuel is passed through the tubes and the oil is guided around the outside of the tubes by baffles in a number of passes.

The oil pressure in the FOHE is always higher than the fuel pressure. This ensures that fuel does not pass into the oil system and then hot regions of the engine – which would be a potential fire hazard.

Low-pressure filter

The purpose of the main LP filter is to protect the downstream units from contaminants and ice in the fuel. The EEC continuously monitors the differential pressure signal across the filter element and indicates to the flight deck impending blockage of the main LP filter.

The LP filter is provided with a bypass valve, which opens at a differential pressure substantially above the point where indication of impending blockage is first given. This valve passes unfiltered fuel but ensures no fuel interruption can occur.

High-pressure pump

The HP pump has to provide sufficient fuel flow at pressure over all engine speeds and operating conditions. The HP pump is typically a gear pump consisting of two inter-meshing gears or can be a plunger-type fuel pump. A gear pump is a constant displacement pump: for each revolution, a fixed volume of fuel is delivered equivalent to the gear tooth volume; therefore, the volume of flow delivered per revolution is constant. The HP pump output pressure is dependent on the HP system backpressure, which is the sum of all downstream unit pressure losses plus the combustor internal pressure.

The gear pump always delivers excess flow relative to the demand, with the surplus fuel spilled back into the LP fuel system. The gear pump dissipates power in the form of heat

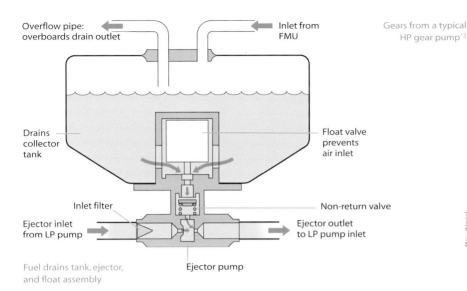

Overflow pipe: overboards drain outlet

Inlet from FMU

Drains collector tank

Float valve prevents air inlet

Inlet filter

Non-return valve

Ejector inlet from LP pump

Ejector outlet to LP pump inlet

Ejector pump

Fuel drains tank, ejector, and float assembly

Gears from a typical HP gear pump [3]

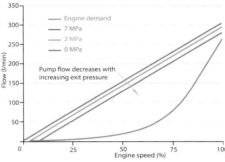

Typical large civil engine HP gear pump flow characteristics at different backpressures

into the spill flow; when engine demand is low, this results in a significant fuel temperature rise. The design aim is to size the gear pump so that it satisfies the highest and lowest engine fuel flow demand while minimising heat input and surplus fuel flow.

Fuel metering unit

A typical FADEC fuel metering unit (FMU) or hydro-mechanical metering unit (HMU) consists of three main valve assemblies:

〉 a spill valve

〉 a metering valve

〉 a pressure raising/shut-off valve.

A FADEC has to achieve the following functions:

〉 fuel flow metering

〉 minimum pressure rise

〉 fuel shut-off

〉 overspeed shut-off

〉 manifold draining on shutdown

〉 pump unloading

〉 HP compressor airflow actuator control

〉 Fuel return-to-tank (FRTT) control.

Valve movement in the FMU is achieved by applying fuel pressure, achieved with the electro hydraulic servo valve (EHSV), or 'torque motor'. The EEC commands the EHSV position for the required valve movement to ensure the desired fuel flow or servo valve actuation.

Fuel flow metering

The engine consumed flow is controlled by the FMU in response to inputs from the EEC, which in return receives a thrust demand via power lever in the cockpit. The EEC trims fuel flow in accordance to the thrust demand, which is translated into a position demand of the metering valve in the FMU. Metering valve position feedback to the EEC is provided by means of a linear, variable displacement transducer. To achieve the demanded engine fuel flow, the spill valve arrangement spills excess fuel flow.

Minimum pressure rise

The pressure raising shut-off valve (PRSOV), often known as the high-pressure, shut-off valve, is located downstream of the fuel metering valve. The PRSOV has several functions, one of which is to maintain a minimum HP pump pressure rise at low flows, so ensuring that there is sufficient pressure available within the FMU at lower engine settings for servo-powered systems (for example, fuel-driven actuators) and to move the FMU internal valves.

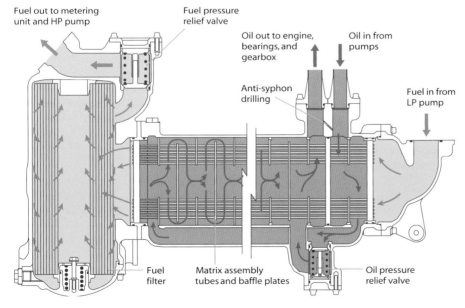

Fuel out to metering unit and HP pump

Fuel pressure relief valve

Oil out to engine, bearings, and gearbox

Oil in from pumps

Anti-syphon drilling

Fuel in from LP pump

Fuel filter

Matrix assembly tubes and baffle plates

Oil pressure relief valve

An example of an LP filter housing bolted to the FOHE

Fuel shut-off

A further function of the PRSOV is to shut off the engine fuel supply. The PRSOV is moved in response to one of three situations:

› cockpit fuel shut-off command

› a command by the engine independent overspeed protection system

› a command by the EEC itself.

Overspeed shut-off

Aero gas turbines must have a shutdown system that is independent of the EEC to allow the engine to be shut down in case of rotor overspeed. This system is known as the independent overspeed protection (IOP) and uses the speed signals from the engine LP and HP shaft speed probes to detect potential overspeed, and to shut off the engine fuel supply via the PRSOV.

Small engines are equipped with a mechanical actuation system as only these systems accomplish the required fast reaction time. Larger engines do not need as quick a response as the turbine acceleration is slower due to the substantially higher moment of inertia of the rotating assemblies involved; this means they can use electronic systems rather than the quicker but more complex mechanical systems.

Manifold draining

When a gas turbine is shut down, the fuel drains system uses combustion chamber pressure and gravity to purge the fuel contained in the fuel manifold and feed pipes into the drains tank. This prevents fuel lacquering in the fuel spray nozzles as a result of heat soak from the hot combustor; it also ensures compliance to environmental regulations preventing emissions caused by fuel venting into the atmosphere.

Pump unloading

Opening the spill valve to recirculate the HP pump delivered flow provides pump unloading. This is necessary because the engine fuel system becomes dead-headed during shutdown – there is nowhere for any remaining fuel flow to pass, which could cause a build-up of pressure. Unloading also reduces parasitical mechanical drag on the accessory gearbox during engine in-flight windmilling.

HP compressor airflow control

An actuation system, comprising fuel-driven hydraulic actuators plus a system of links, is used to adjust the variable stator vanes (VSVs) in the HP compressor to prevent compressor stall and surge (» 96 – 99).

Fuel return-to-tank control

For heat management purposes, heated fuel from the engine may be returned back to the aircraft tank. The corresponding control valve returns excess fuel from the HP pump to the aircraft tank in response to EEC command or pilot input.

Fuel flow transmitter

The flowmeter is a mass flow measurement device. The EEC provides a conditioned flowmeter signal to the flight deck where it is used for aircraft purposes only (for example,

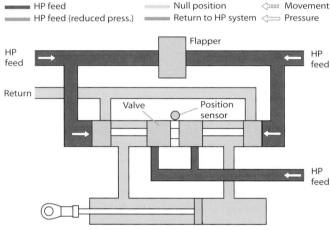

Null condition - flapper centralised, no flow to actuator, pressure equal

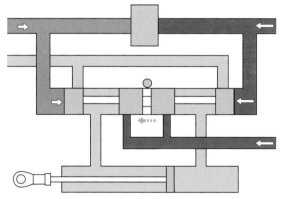

Flapper moved to left, pressure imbalance, valve starts to move left

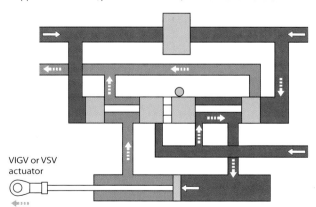

Valve has moved, flow to actuator: VIGVs/VSVs are actuated.
Flapper returns to null position.

Electro hydraulic servo valve for fuel actuation of the VIGVs and VSVs fuel flow indication on the flight deck and input into the aircraft flight management system); the signal is not used for any engine control function.

HP fuel filter
The HP filter is a relatively simple unit that provides a final protection for the set of fuel injectors. The presence of an HP filter is also a certification requirement.

Fuel manifold and fuel spray nozzles
The fuel burner manifold feeds fuel to the spray nozzles, or fuel injectors, in the combustor. The fuel injectors atomise the fuel in the combustion chamber where it is burned, releasing heat energy.

On large engines, the correspondingly large diameter of the combustor causes variation in the fuel pressure head due to the height difference between the top and bottom of the manifold. This pressure head can result in uneven distribution of fuel through all injectors, and a time delay between the flow of fuel from injectors at the bottom and those at the top of the combustor. To overcome this problem on large engines, weight distributors are used. However, on small engines, this is typically overcome by feeding fuel into the manifold at the midpoint.

The weight distributors use a valve, spring, and mass design. The fuel pressure at the fuel injectors located at the bottom of the combustor must overcome both a spring force and also the force due to the mass – whereas the fuel pressure at the fuel injectors located at the top of the combustor need only overcome a reduced spring force since the mass is acting on the spring. This approach ensures that all fuel injectors are primed with fuel and that the fuel flow into the combustor is uniform around the combustor during engine start.

The typical type of fuel injector used on modern engines is an airspray injector. Rather than relying solely on the pressure drop over the injector to atomise the fuel, this type of injector uses combustion airflow to aerate the fuel. This approach reduces the required fuel pressure, and subsequently the fuel system pressure, allowing a gear-type pump to be used in the fuel system (» 120 – 121).

Heat management
Heat management is the process of ensuring the optimum use of heat generation and rejection to maintain the oil system and fuel system temperatures within their respective temperature operational limits, while ensuring minimum engine performance losses.

Oil is used to lubricate and cool electrical generators, bearings, and gears in the transmission system. This results in a large amount of heat transferred to the oil; in order to maintain the oil (and the components that the oil is cooling) at acceptable temperatures, it is necessary to remove this heat. This transfer of heat to the oil represents an energy loss from the engine thermodynamic cycle and, if this heat is lost permanently, it can cause a significant performance penalty to the engine. One convenient way of recovering this heat back into the engine thermodynamic cycle is to dissipate the energy into the engine fuel flow. This also has the advantage of heating the fuel to prevent exposure of the fuel system components to fuel-borne ice particles.

The starting point for designing the heat management system is to consider using the engine fuel flow alone as the heat sink – as this will give the simplest possible system configuration. If the engine fuel flow heat sink is insufficient, then it is necessary to incorporate supplementary cooling or change the heat rejection levels to limit the exposure to minimum and maximum temperatures.

To aid the design of an optimum heat management system, a fully comprehensive heat management model is used to determine the heat generation in each aspect of the oil and fuel system.

Oil temperature control
The main heat management issue for oil is to limit the maximum oil temperature. If the oil temperature is too high, the properties of the oil may be inadequate – and if the oil temperature is excessive, the oil may degrade resulting in loss of properties, formation of solid particulates, and possibly auto ignition that could ultimately lead to an engine fire.

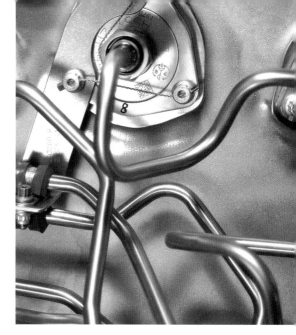

High fuel temperature control
High fuel temperature can result in thermal degradation of the fuel, thereby producing lacquer that can cause problems with fuel system components. High fuel temperatures are mainly a concern at low fuel flow rates, such as idle, and especially during transient operation of the engine when the engine power is reduced. During deceleration of the engine from high power to low power, the fuel flow rapidly reduces, but the oil temperature does not respond as quickly. This results in the heat generation of the oil system at high power being transferred into a low fuel flow. The fuel temperatures in this circumstance could be excessive, which is why a suitable heat management system is required.

Low fuel temperature control
A responsibility of the heat management system is to ensure that the fuel temperature is above $0^{\circ}C$ at the critical and vulnerable parts of the fuel system for the majority of operating conditions. The heat rejected by the engine oil system via the FOHE is normally sufficient to ensure that fuel temperatures are above $0^{\circ}C$, so protecting the fuel system from fuel-borne ice. However, there can be a deficit in the heat provided to the fuel resulting in fuel temperature below $0^{\circ}C$ when operating from idle to a high engine power setting. When high thrust is demanded, the engine fuel flow increases rapidly, but the oil system temperatures do not increase at a similar rate. This means that the heat generation of the oil system at low power is transferred into a very high fuel flow. In these circumstances, the fuel temperatures could reduce below $0^{\circ}C$, leading to water in the fuel freezing and becoming ice. As the engine power setting

is maintained, the engine oil system heat generation increases, and subsequently the heat transferred to the fuel increases, resulting in fuel temperatures increasing.

If this circumstance cannot be avoided by the design of the heat management system, the fuel filter is used to collect the ice while the fuel temperature is below $0°C$.

Gas turbine fuels

The two main fuels used for gas turbines are kerosene (essentially a paraffin) and 'wide cut'. Kerosene fuels have improved safety for handling compared to wide cut and gasoline types, and is the fuel used predominately in civil aviation. Wide cut fuels provide a higher yield of product per unit volume of crude. This is an advantage in certain supply scenarios where fuels are in short supply. In general, the use of wide cut types is becoming increasingly rare. A fuel used less frequently is high flash fuel, which has improved safety for handling in confined spaces and is used on, for example, aircraft carriers.

Additives are used to enhance specific aspects of the fuel performance:

› Fuel system icing inhibitor reduces the risk of fuel system or LP fuel filter blockage from ice.

› Corrosion inhibitor/lubricity aid improves fuel lubricity, which can reduce fuel pump and component wear.

Fuel properties such as density and viscosity impact upon the flow region of the fuel; this affects the pressures in the system and the heat transfer to the fuel. The ability of the fuel to absorb heat is dependent on the specific heat capacity of the fuel. Fuel contains dissolved water (approximately 0.028 per cent by volume), which at low fuel temperatures will separate from the fuel and freeze, potentially blocking fuel system components. This is a particular concern for operation on very cold days ($-54°C$). Fuel contains dissolved air from the atmosphere (approximately 15 per cent by volume compared to three per cent by volume for water). This dissolved air is not normally a concern for the fuel system due to sufficient fuel pressure, but if fuel pressure is low then the dissolved air will come out of solution. Furthermore, fuel turns

to vapour when exposed to pressure below the vaporisation pressure, or when operating at high fuel temperatures particularly during operation on very hot days ($55°C$). The result is an air and fuel vapour mix that can adversely affect the delivery of fuel flow. Wide cut fuels are more susceptible than kerosene to fuel vaporisation due to their more volatile nature.

The oil system

All aero gas turbine engines incorporate an oil system to provide lubrication, cooling, and corrosion protection for gears, bearings, and splined shaft couplings. Oil may also be used as a sealing medium between rotating shafts. The oil system is an important element in the monitoring of engine health.

A successful oil system ensures satisfactory engine operation and a long service life. Specialised lubricants allow operation over a wide range of temperatures, pressures, and engine speeds.

Turboprop engines incorporate additional oil system features required by the heavily-loaded propeller reduction gears and propeller pitch control mechanism.

Most gas turbine engines use a self-contained recirculatory oil system that distributes oil to components throughout the engine; the oil is returned to an oil tank by pumps. The oil must be cooled to prevent overheating and loss of oil properties. Air or fuel is used for this purpose. Heat from the oil is generally used to prevent ice formation in the fuel system.

The oil must maintain its properties through the service life of the engine, as it is not normal practice to change the engine oil during routine service.

Oil system description

The engine oil system is constructed from three complementary sections:

› a pressure feed and distribution system

› a scavenge system

› a vent system.

There are two basic forms of recirculatory systems: the full flow and the pressure relief valve system. The major difference is in the

control of the oil flow. In all current engine designs, the oil pumps are powered by a geared drive from the highest speed engine main shaft. Satisfactory operation is critical to the safe operation of the engine. Oil temperature and oil pressure are indicated on the flight deck.

Operation of a typical system

Oil from the tank is drawn through a strainer (to protect the pump from any contaminant in the tank) to provide a supply of pressurised oil. The oil then passes through a filter to a pressure-limiting valve. This protects against excessive pressures caused by a blockage or highly viscous oil during very cold starts. A pressure relief valve system also has an operating pressure control valve at this point.

The oil then flows to the heat exchangers before being separated into individual lines to supply each bearing chamber and the gearbox. Jets and distributors meter and direct the flow as required. In a turboprop, engine oil is also supplied to the propeller pitch control system, reduction gear, and torquemeter system. Having performed its lubricating and cooling task, the oil is directed to a sump – there are separate sumps for the gearbox and each bearing chamber. Scavenge pumps, again protected by strainers, extract this oil and return it to the tank via the scavenge filter.

On entering the tank, the oil is de-aerated ready for recirculation. Separated air from the scavenge and vent systems is exhausted overboard through the breather.

Full flow system

Most modern oil systems use a full flow arrangement, which allows smaller oil pumps to be employed than an equivalent pressure relief valve system.

The full flow system is also more able to approach optimum oil flow rates throughout the engine speed range. Full supply pump delivery flow is delivered to the oil feed jets. This system uses the full capacity of the pumps at the maximum speed. Restrictors at the end of each branch of the system determine the distribution of flow. A disadvantage of this system is that, if the bearing chambers are unequally pressurised, the proportion of the

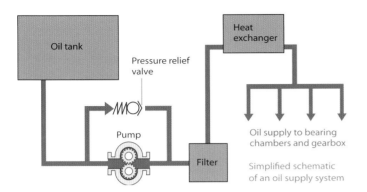

Oil tank

Pressure relief valve

Heat exchanger

Pump

Filter

Oil supply to bearing chambers and gearbox

Simplified schematic of an oil supply system

total flow received by each bearing chamber may vary through the speed range of the engine. This can impact on scavenge pump sizing. With a full flow system, the indicated oil pressure will change according to the engine operating condition. Pressure relief valves may be used to protect system components from the extreme pressures that could be generated in abnormal circumstances.

Pressure relief valve system

In a pressure relief valve system, the oil flow to the bearing chambers is controlled by limiting the pressure in the feed line to a given design value. Typically, this is achieved at idle, giving a constant feed pressure over normal engine operating speeds. A spring-loaded valve allows surplus oil to be returned from the pressure pump outlet

to the oil tank, or pressure pump inlet, when the design pressure level is exceeded. The spilled oil flow represents overcapacity in the pumps, hence their larger size compared to those in a full-flow design. This system suits engines that have low levels of bearing chamber pressurisation. Many engines have bearing chamber pressures that rise sharply with increasing power, reducing the pressure difference between the bearing chamber and oil supply pressure. The oil flow rate to the bearings then reduces as engine speed increases. To alleviate this problem, the increasing bearing chamber pressure may be used to augment the relief valve spring load (pressure-backed relief valve systems). This gives constant flow at higher engine speeds by increasing the pressure in the feed line as the bearing chamber pressure increases.

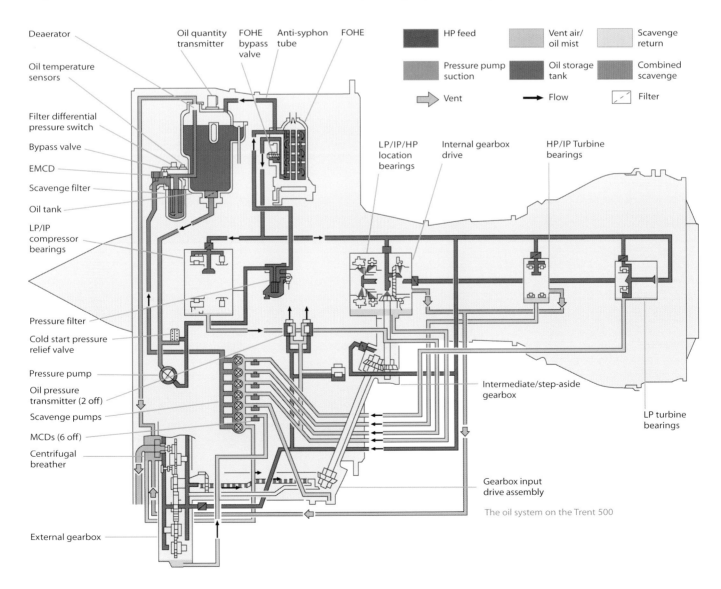

The oil system on the Trent 500

Scavenge system

The oil supplied to the bearing chambers must be evacuated and returned to the tank as quickly as possible. This minimises oil 'gulp' and exposure of the oil to high temperatures while also maximising the useable oil tank contents. Each bearing chamber will normally have a dedicated pump, as will the gearbox. Flows from these pumps are combined and returned to the tank in a single pipe. This flow is a mixture of oil and sealing system air. A de-aerator in the oil tank separates the oil from the air. The air is then vented through the breather.

The combined scavenge line is normally filtered and contains a master chip detector. Provision is usually made so that, if required for diagnostic purposes, chip detectors can be fitted in each individual scavenge line.

The temperature of the combined scavenge oil is often used as the primary indication of oil system temperature on the flight deck.

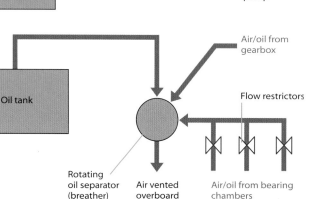

Simplified schematic of a scavenge system

Simplified schematic of a vent system

Vent system

It is essential to prevent oil leakage from the bearing chambers. To achieve this, pressurised seals are used. To ensure that the pressure drop is always into the bearing chamber, it is normal practice (for chambers with more than one oil seal) to provide a vent to a lower pressure. The vent system capacity is sized to ensure that the sealing airflow used is sufficient to ensure bearing chamber sealing, with minimal impact to engine performance. The air vented from the bearing chambers contains oil, which must be separated and retained in the system. The air is vented overboard. A rotating oil separator (breather) is used to recover the oil from the vent flow.

Oil filtration

There is a direct link between oil cleanliness and the life of components within the oil system. Filtration is used to maintain the oil in a clean condition.

A typical turbofan engine has two stages of oil filtration: primary filtration is provided by a large-capacity filter in the combined scavenge line to remove fine particles; a second filter, the pressure filter, is provided after the oil feed pump. The pressure filter is much coarser than the scavenge filter and ensures that the small

passages in the coolers and oil jets are not at risk of blockage from particles when the fine scavenge filter is bypassed. The pressure filter does not have a bypass, and if it were to block a low oil pressure warning would be generated. The oil system is therefore always protected by a degree of filtration.

Oil system differences for marine applications

A typical marine gas turbine installation may consist of a gas turbine change unit (GTCU), a power turbine, and the associated installation module. The GTCU oil system may share oil with a hydraulic system pressurised by a GTCU-driven pump. The hydraulic system provides power to the airflow control regulator.

Module-mounted components

Some oil system components, such as the oil tank, may be located in the installation module for ease of access. Other components, which do not require regular attention (for example, the engine oil-pumping unit), may be located on the high-speed gearbox on the GTCU, as on an aero application.

The main filtration is carried out in the scavenge side of the system and the filter

may be a duplex unit, located in the installation module. This allows the oil flow to be switched between two identical filter elements, allowing one to be replaced without stopping the engine or losing filtration. If a filter becomes blocked a bypass valve will open, allowing unfiltered oil to flow permitting the engine to continue running with no loss of oil pressure. A visual indication of the filter condition (pressure drop) is provided to ensure that the filter is changed before bypass occurs. Instead of using the fuel (or air) supply to cool the oil, it is usual to pass sea water through a module-mounted heat exchanger.

Power turbine oil system

The power turbine oil supply system is independent of the GTCU system. Oil is delivered from the ship's supply system, through an adjustable orifice valve and distribution block, to the power turbine bearings; the oil is returned to the same supply system.

Oil system components
Distribution system

The distribution system is used to feed oil to components such as bearings, gears, seals, and splines. A branched pipeline system passes oil to the bearing chambers and gearbox. Oil is transported across the gas path in pipes inside hollow vanes. Oil jets at the end of the pipes are used to meter the flow and direct oil to the components or into rotating distributors. These distributors ensure the correct proportion of the oil supply is provided to each location.

Oil tank

The oil tank provides a reservoir of oil to supply the oil system, either as a separate unit or as an integral part of the external gearbox. It must have provision for draining and replenishment. An electrical quantity transmitter and sight glass are usually incorporated to monitor the oil contents. Filling is by either gravity or pressure connection. Engines designed to operate for extended periods in zero or negative gravity flight conditions will have oil tanks that incorporate features ensuring a continuous supply of oil. Turboprops require a separate reservoir of oil that cannot be depleted by leakage from the basic oil system so that that the propeller pitch can be feathered if the engine has to be shut down in flight after loss of oil. A de-aerating device is incorporated within the oil tank to remove the air from the returning scavenge oil. The capacity of the tank must be sufficient for the longest flight to be undertaken with the maximum allowable oil consumption. The tank design must also accommodate temperature-related expansion of the oil.

Anti-siphon precautions

There is potential for oil in the system to siphon from high level to lower levels when the engine is not running, resulting in delayed oil supply when the engine is next started. Connecting the highest point in the feed system pipework to the oil tank breaks the siphon that creates this effect.

Oil feed pump

The oil system feed pump is typically of the vane, gear, or gerotor type. These are positive displacement pumps that deliver a known flow, proportional to pump speed. The oil pressure is generated by the resistance to the oil flow in the pipe backed by the bearing chamber pressures. The pump may incorporate an anti-drain valve to prevent oil leaking from the tank to the gearbox

A typical marine oil system

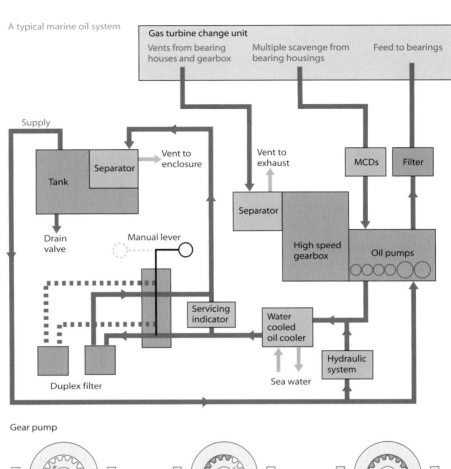

Gear pump

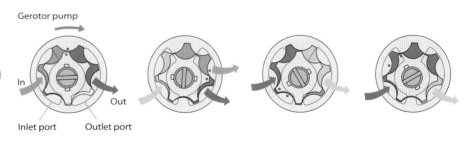

Gerotor pump

Vane pump

Three types of oil feed pump: gerotor, gear, and vane

183

Pressure filter

A typical thread filter acts as a coarse strainer between the tank and oil pumps

while the engine is not running. Turboprops may use an additional pump to supply oil to a torquemeter of the propeller pitch control mechanism.

Pressure filter

The pressure filter is sited after the oil feed pump. Usually, it is relatively coarse (125 micron) and does not have a bypass.

Strainers

Coarse strainers are usually fitted at the outlet from the oil tank or immediately prior to the inlet to the oil pumps to prevent any extraneous material from damaging the pumps. Thread-type filters are often fitted as a 'last chance' filter immediately upstream of the oil jets. Sometimes perforated plates or gauze filters are used for this purpose and for protecting the pumps in the scavenge system.

Oil cooling

The circulating oil acquires a large amount of heat. To maintain the oil at acceptable temperatures, this heat must be removed. Heat exchangers, situated in the feed or scavenge systems, transfer oil heat to the fuel or to the air. Significant sfc losses can result from poor management of this heat transfer.

Exposure to very low ambient temperatures while the engine is shut down can result in highly viscous oil in the heat exchanger matrix. The flow of cold fuel or air through the heat exchanger matrix after starting keeps the oil cold and oil circulation may be inhibited.

In these circumstances, a pressure-operated bypass valve opens, allowing oil through a small portion of the matrix, limiting the pressure drop and heating the matrix to establish full flow.

Fuel oil heat exchanger

The FOHE transfers oil heat to the fuel, and is typically situated within either the high-pressure oil feed system or the lower pressure scavenge system. One concern with such heat exchangers is that fuel may leak into the oil and this combustible mix could then be passed to the hot bearing chambers and components. To avoid this situation, the FOHE is typically located in the high-pressure oil feed system, and the oil pressure is maintained above fuel pressure to prevent leakage of fuel into the oil. The FOHE is usually positioned upstream of the fuel filter to allow the heat from the oil to keep the fuel filter free of ice.

Air oil heat exchanger (AOHE)

Air cooling must be kept to the absolute minimum in order to reduce performance penalties. Attention is given to achieving low pressure losses in the air system to give maximum pressure drop across the exhaust nozzle, so improving thrust recovery.

Oil jets

The desired flow of oil to a component can be achieved by use of a suitably sized restriction at the end of the oil line, known as an oil jet. The design of the jet can provide either a spray or a targeted, coherent stream of oil, directed to a component or to a catching feature that will then feed the component. To achieve low oil flows without using unacceptably small jets, multiple restrictors may be used upstream of the jet to reduce the final jet pressure drop.

Oil distributors

Some components are not readily accessible to an oil jet; in this situation, distributors are used, employing centrifugal forces to distribute the oil. These devices fit within rotating shafts and are supplied with oil from a jet. Features in the bore of the distributor segregate the supplied flow into discrete flowpaths, each of which has an exit through the shaft at the appropriate point to lubricate the component. The outlets may be positioned at almost any point along a shaft, and the shaft may have any orientation, as the effects of rotation ensure that the oil reaches the outlet point.

Starting oil troughs

Gears in the engine starting drive system are heavily loaded early in the start cycle, before the oil system is able to supply a pressurised flow of oil. To provide some oil for the first seconds of starting, a trough may be provided to collect and retain some oil after shutdown. The gear to be lubricated sits in this pool of oil and so has some lubrication during initial rotation.

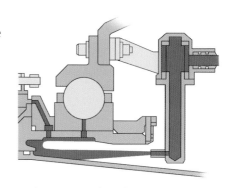

A schematic view of an oil jet

Oil jets can provide sprays or targeted streams of oil

Bearings

The quality of the oil presented to bearings is particularly important. Solid contaminants can cause damage; therefore, filtration of the oil is vital. Magnetic chip detectors are used in the scavenge system to collect steel debris and so detect deterioration of the bearings before a failure occurs. Regular chip detector inspection reduces the risk of an unexpected bearing failure.

Spline lubrication

Without lubrication, the articulation of splines used to connect shafts can lead to wear. Several methods of lubricating the splines are in use:

❭ Grease packing – the spline is packed with grease on assembly and an o-ring retains the grease in the spline.

❭ Oil splash/mist – an oil/air mist flow is induced through the spline.

❭ One shot lubrication – a quantity of oil is put into the splines on engine starting or shutdown.

❭ Dedicated lubrication – a continuous flow of oil can be provided throughout engine operation.

The oil distributor uses centrifugal force to cause oil to flow from the central oil tube

Left: Magnetic chip detector – inspected manually

Above: An electronic magnetic chip detector – removes need for manual inspection

electronic chip detector in the master position to provide earlier notification of an impending problem.

Vent system
Bearing chamber sealing
Sealing against oil loss between rotating shafts and bearing chambers is an important feature of engine design. Oil leakage can lead to severe out-of-balance of the rotating assemblies causing vibration, and is also a fire risk. Oil leakage into the compressors and cabin air off-take system can be detrimental to air quality. An appropriate sealing method is selected from several options available (labyrinth seal, oil-backed labyrinth seal, carbon ring seal, hydraulic seal, brush seal, and metal ring seal).

Breather
Air vented from the bearing chambers, gearboxes, and oil tank is exhausted overboard through a breather. The vent airflow will normally contain oil droplets. It is undesirable for this oil to be lost from the system, as it would contaminate the environment. A centrifuge, rotating at high speed, achieves separation of the oil droplets. Separated oil is returned to the oil tank by the scavenge system, leaving clean air to be ejected overboard.

The design challenge
Aero gas turbine oil systems must be reliable, lightweight, and cost-effective. They must maintain acceptable lubricant and system component operating conditions at all times. The impact on fuel temperature must be beneficial at low temperatures and acceptable at high temperatures. Any negative impact on fuel efficiency must be minimised and oil consumption must be low.

These requirements are increasingly difficult to achieve as engine designs become more efficient. Factors working against the oil system include increased shaft speeds, contra rotation, increased pressures and temperatures, reduced specific fuel flows, and reduced space available for bearing chambers.

Scavenge oil system
Oil scavenge pumps
Scavenge pumps generally follow the same construction as the oil feed pump. Each bearing chamber or gearbox is serviced by a dedicated scavenge pump, except where bearing chamber pressure or gravity can be used to drive the oil to a shared sump. The capacity of a scavenge pump is usually much greater than the oil flow it is required to return to the tank. This accommodates non-linear flow/speed relationships, and aeration of the oil. It is usual to protect the pumps with a strainer at each inlet.

Scavenge filter
As described earlier, the primary filtration of the engine oil is provided by a large capacity filter, located immediately upstream of the oil tank in the combined scavenge line. This primary filter has a three micron rating.

Magnetic chip detectors
Provision for removable magnetic plugs is provided in all scavenge lines. The plugs have a magnetic probe positioned in the oil path. If a bearing deteriorates, any material that is released is caught on the probe. The material can be analysed to indicate which component is wearing.

A master chip detector is positioned in the combined scavenge line, upstream of the scavenge filter. This probe is always fitted, and is routinely inspected. If material is found on this probe, the others may then be fitted and inspected to identify the source of the material. Modern engines may use an

Ensuring oil system integrity

Engine testing is the primary way of ensuring the integrity of the oil system, backed up by computer modelling and analysis. Testing is carried out in sea-level test cells or on test aircraft. Specific component testing is used where appropriate. Component tests cover

> the fireproof capabilities of the system components

> oil pump performance and durability

> heat exchanger performance and durability.

Engine tests cover

> usable oil tank contents

> component integrity following fan blade release

> starting under extreme low temperatures (minimum oil temperature, maximum oil pressure)

> maximum oil temperature

> minimum oil pressure

> windmilling operation

> oil flow interruption.

Oil system health monitoring

Engine health monitoring is an essential aspect of the successful and cost-effective operation of modern gas turbine oil systems. Key parameters are recorded and monitored:

> engine oil pressure (differential)

> engine oil temperature

> HP filter pressure drop

> scavenge oil filter pressure drop

> Oil level in the tank.

The importance of engine health monitoring in reducing life-cycle costs on future engines will lead to a more comprehensive sensor list, which will employ new technology and analysis systems.

Lubricating oils

The development of turbine oils has made a significant contribution to the success and reliability of gas turbines. Early engines ran on mineral-based oils with no additive technology. Practical synthetic oils became available in 1947.

The first synthetic gas turbine oils were based on select organic esters, which had been chemically produced from naturally occurring materials, and some petrochemical derivatives. The initial simple diesters had a viscosity of approximately three centistokes ($3m^2/s$) at 100°C and were suitable for the military turbojets of the day. However, these oils were not suitable for the heavily-loaded gearboxes used in turboprops. Thickeners were added to the base oil raising the viscosity to 7.5 centistokes at 100°C and improving the load-carrying performance of the oil at a cost to the low temperature fluidity.

Continuing improvements in engine performance resulted in increasing rates of heat rejection to the oil and higher engine operating temperatures. These developments drove the diester-based oils towards their limits of performance. Oils may experience a temperature range between -40°C to 250°C in engine operation. Advances to the original diester fluids resulted in polyol ester-based oils being developed during the early 1960s. These oils (5 centistokes at 100°C) offered a general improvement in performance, and are widely used in today's turbofans.

The rotating oil separator, or breather, separates oil and air, recovering oil droplets to the oil system that could be vented overboard

At this point the engine has everything it needs to run.
But it doesn't know what to do.

control systems

THE CONTROL SYSTEM MONITORS AND CONTROLS ALMOST EVERY ASPECT OF A GAS TURBINE. EVERY SECOND, IT CAN TAKE 200 MEASUREMENTS FROM MORE THAN 40 SENSORS; THE CONTINUOUS ADJUSTMENTS IT MAKES TO THE ENGINE OPERATION ARE ESSENTIAL TO SAFETY AND EFFICIENCY. WITHOUT IT, THE MODERN ENGINE WOULD NOT FUNCTION.

control systems

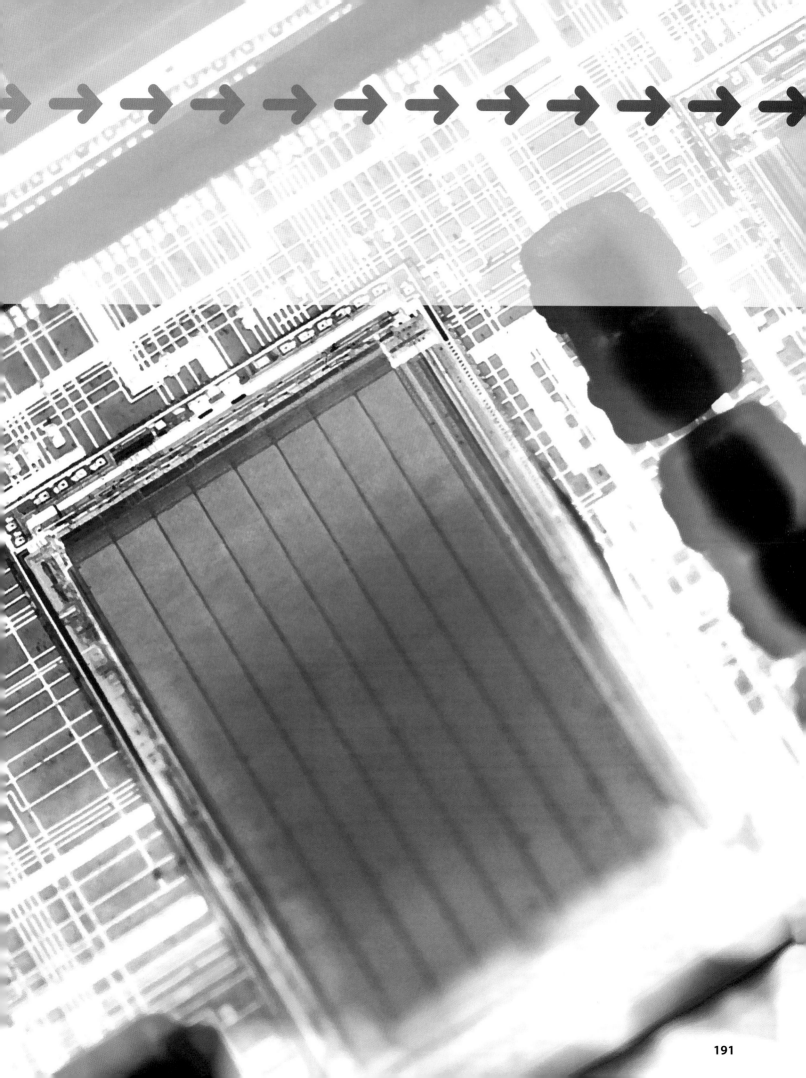

A control system is designed to remove, as far as possible, workload from the pilot or operator, while still allowing him or her ultimate control of the engine. To achieve this, the control system monitors inputs such as

> shaft speeds

> engine temperatures

> oil pressures

> actuator positions

and, when the operator selects a power setting, the system then sets a range of variables:

> fuel flow

> variable stator vanes

> air bleed valves.

When a change of thrust is required, the control system ensures that all these variables are adjusted in order to achieve the desired thrust efficiently while maintaining the engine safely within its operating limits.

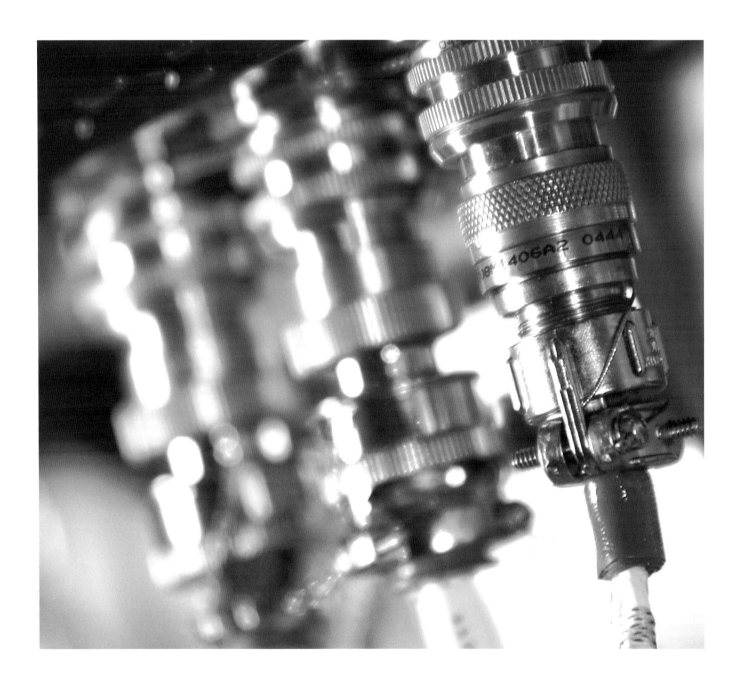

The gas turbine engine has many different applications, each with different requirements, and therefore each with its own control equipment and system architecture. However, the basic principles and functions of a gas turbine control system are essentially the same for all applications.

Principles and functions of a control system

After initial checks, the control system is required to start the engine, and accelerate it safely to a point where the gas turbine can sustain its speed without starter power and is stable (idle speed). Thereafter, the pilot or operator will require various levels of power output, depending on the operation required. The control system accelerates or decelerates the engine by changing the fuel flow and manipulating compressor variables (and others) to ensure the manoeuvres are smooth and surge-free. During deceleration, care must be taken not to reduce fuel flow below the point at which combustion would be extinguished. When the pilot or operator shuts down the engine, the controller sets fuel flow to zero, and the engine decelerates to a stop. In some applications, further tasks are carried out to ensure that maintenance on the engine can be carried out safely and the engine is prepared for the next start. Before, during, and after the operation of the engine, data is transmitted by the control system for display to the operator (» 253).

Expressed in these terms, the control system's task is simple, but there are some additional complexities. For example, determining the engine power required by the pilot or operator involves a rating calculation, which, in an aero-engine application, involves flight condition (altitude and Mach number) and takes into account the non-propulsive power being extracted for aircraft services. Thus, for a given nominal power demand from the aircraft during climb, actual power will be varying continually.

The control system also has to perform self-checks; it ensures it is operating without failures and it must not be working with incorrect data – either situation would result in erroneous control decisions or incorrect data being sent to the pilot or operator. The rigour of the design and analysis of the control system reflects the safety, economic, and other consequences of such an error.

Above all, the control system must ensure that the engine is operating safely within its defined limits, even if the engine or control system fails. In some circumstances, the control system has no alternative but to shut down the engine – for instance, if there is a danger of rotor overspeed because the electronics can no longer control the flow of fuel to the engine. There is nothing the electronics can do in these circumstances and the rate of change of fuel flow may be too rapid to expect the operator to intervene. For this reason, all systems have independent means of measuring a limited set of data (typically rotor speeds) and commanding an immediate engine shutdown, or some other safe state, if set limits are exceeded. The control system contains many features designed to provide this safety protection, and the design and testing of these features is a major part of the designer's task.

Control laws

Each manufacturer has different control strategies, and each engine type has detailed differences in its control laws. However, aerospace applications place certain common requirements on control:

> An engine must be able to accelerate from low power to high power in a fixed time so that an aircraft can abort a landing and achieve max take-off thrust – for example, to avoid a runway obstruction. The control laws may use a closed loop acceleration algorithm, where rate of change of speed is a function of current speed, to ensure that at a given condition acceleration time is always the same.

> As an engine wears during its life, the thrust it provides at a given condition must remain above a certain level if the aircraft is to achieve its take-off performance. Thus, a parameter must be chosen which provides a close measurement of thrust, and any inaccuracy in the measurement compensated by providing additional power. The control system must then control to that parameter very accurately.

> An engine must accelerate from stationary to idle in a reasonable time in order that the aircraft can taxi under its own power. The starting algorithms must accelerate the engine at a rapid rate, avoiding any stall or stagnation regions.

> The pilot must always be able to shut the engine down – the system's hardware must provide a separate mechanism to allow the pilot to override the control system if required.

> Above all, an engine must always be operated within its safe limits. The control system, therefore, must be programmed with data on all the relevant limitations and the action to be taken if such a limitation is approached.

This is necessarily only a small subset of the engine control requirements and consequences on the system.

Engine left hand view

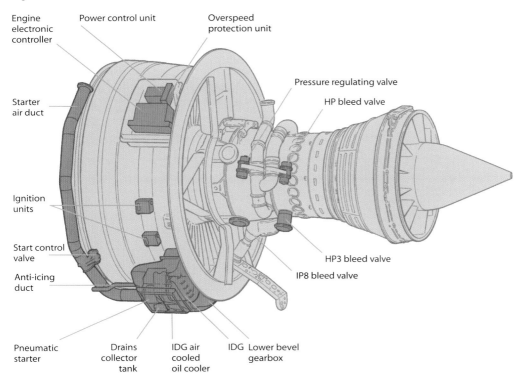

Engine
electronic
controller

Power control unit

Overspeed
protection unit

Starter
air duct

Pressure regulating valve

HP bleed valve

Ignition
units

Start control
valve

Anti-icing
duct

HP3 bleed valve

IP8 bleed valve

Pneumatic
starter

Drains
collector
tank

IDG air
cooled
oil cooler

IDG Lower bevel
gearbox

Engine right hand view

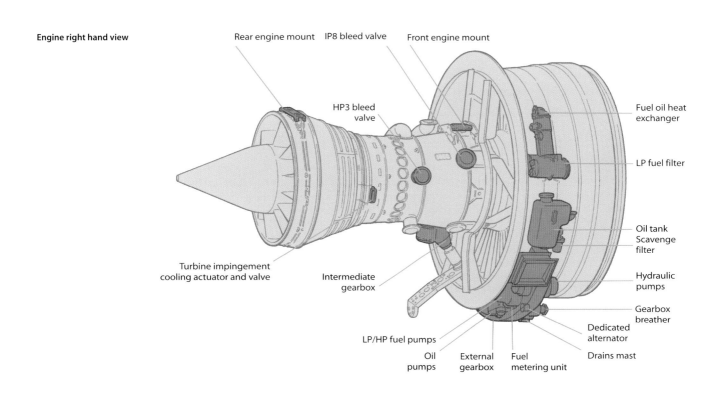

Rear engine mount

IP8 bleed valve

Front engine mount

HP3 bleed
valve

Fuel oil heat
exchanger

LP fuel filter

Oil tank
Scavenge
filter

Hydraulic
pumps

Turbine impingement
cooling actuator and valve

Intermediate
gearbox

Gearbox
breather

Dedicated
alternator

Drains mast

LP/HP fuel pumps

Oil
pumps

External
gearbox

Fuel
metering unit

195

Components of a control system

The complex functions described above are performed most effectively by digital electronics. All modern engines feature this form of control, and many older engine designs have been modified to include it. However, there are some purely mechanical control systems in service.

Control systems for aerospace (and some marine) applications often use bespoke electronic and mechanical equipment because these applications have limited space for their systems, which must also be low in weight. Energy and other marine applications do not have the same restrictions so their control systems can be implemented using equipment closer to industrial standards.

A typical engine control system has many constituents:

❯ An electronic controller that computes and commands the control functions; it contains one or more microprocessors and other circuitry, which read data from sensors, and control actuators and valves.

❯ Engine parameter sensors, including pilot power demand and feedback signals from actuators.

❯ Fuel pumps.

Three types of system architecture: federated, centralised, and distributed

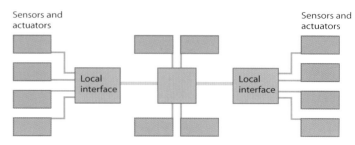

Distributed architecture

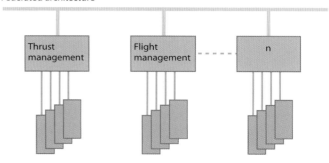

Federated architecture

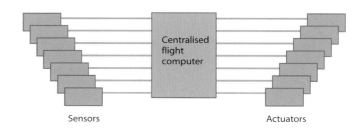

Centralised architecture

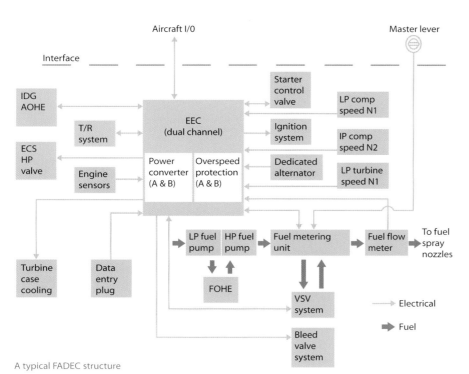

A typical FADEC structure

❯ A means of metering fuel being delivered to the engine, and of shutting the fuel off.

❯ Actuator systems to provide variable geometry control and/or modulation of secondary systems (for example, bleed valves, variable compressor stator vanes and tip clearance control actuators).

❯ An electronic ignition system to generate high voltage sparks at the surface of an igniter plug in the combustion chamber. Under normal circumstances, this is only required to initiate combustion, which is then self sustaining.

❯ A means of controlling the engine starter. The most common form of starter is an air turbine system connected to the accessory gearbox. High-pressure air is used to rotate the HP turbine.

> A means of communication with the vehicle or plant systems. Today, this is usually with an electronic serial databus using an industry standard appropriate to the application, bandwidth, and integrity requirements.

> Separate systems dedicated to ensuring that control system failures cannot result in a dangerous condition.

> The other 'component' of the system is the software in the microprocessor, which has to implement the complex functionality required. There are different standards for the development of this software in different industries.

Civil aircraft engine controls

Controllers for modern engines are based on digital electronics. For historical reasons, the collection of control system elements in an aero engine is often referred to as the Full Authority Digital Electronic Controller (FADEC).

The components of a FADEC system are similar to those described in the general system above, with typically the following additions:

> an engine-driven generator, dedicated to power the FADEC system

> means to control a thrust reverser, if fitted.

Engine electronic controller

At the centre of the FADEC system is the engine electronic controller (EEC). The stringent requirements for safety and availability of an aero engine cannot be met with simplex electronics. For this reason, FADEC designs generally include two channels of electronics, sensors, wiring harnesses, and duplicated electrical parts of actuators, so that the system is fully operational following a single electrical or electronic failure. The two channels within the EEC include features that enable them to exchange data, which is used to detect failures in the system and to allow continued operation. However, the channels must be designed so that a fault in one channel cannot propagate to the other.

In some cases, the two electronic channels are housed in separate enclosures, but more usually, they are contained in a single unit. The EEC may be installed in the airframe, particularly in military aircraft or in civil

A FADEC circuit board[*3]

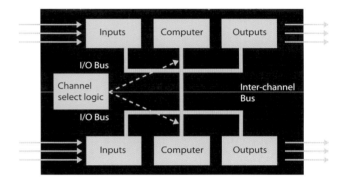

Typical, dual channel EEC arrangement

fuselage-mounted engine installations. For wing-mounted engine applications, typical of large civil turbofans, the EEC is mounted on the engine. This installation places particularly harsh environmental requirements on the electronics, while further emphasising the need for low weight and volume – a need reflected in the components used, the construction techniques, and mounting arrangements. One environmental threat, particular to electronic systems, is electro-magnetic radiation from, for example, lightning (both on the ground and in the air) and airport radar. The substantial connector housings used are in part designed to help alleviate these threats.

The EEC reads data from the sensors, other information from the aircraft avionic systems, and the pilot's inputs to calculate the new required position of the actuators, and uses its drive circuits to move them, often by means of secondary servos in the actuators. It also transmits data relating to the engine condition back to the aircraft, along industry standard serial data busses. The aircraft manufacturer is responsible for deciding which data is displayed to the pilot, subject to certification rules and the engine manufacturer's installation manual.

The EEC gathers information on any failures it has diagnosed within the electronics, the remainder of the FADEC system, or in some cases in the gas turbine itself. This information is transmitted to the aircraft systems, but if the system considers itself to be in a safe configuration and no action is required in flight, the information is often not displayed in the cockpit – it is available to the pilot if required, but is intended for use by maintenance personnel on the ground.

This fault information may also be stored within the EEC itself for retrieval by the ground crew and may include more detail than is transmitted to the aircraft. Should the EEC be removed as a result of a suspected failure, this data is also used to assist in the diagnosis of the fault at the repair base.

Fuel metering unit (FMU)

In a FADEC system, a single unit is dedicated to accepting fuel from the pumping system and uses inputs from the EEC to meter the flow of fuel to the engine. A proportion of the high-pressure fuel supply is used, after appropriate filtering, to power a hydraulic servo system, which operates valves within

Cockpit of a modern aircraft with multi-function displays

the unit (**»** 177–178). One of these valves maintains a constant pressure drop across a port in the sleeve of a second valve. A two-stage servo uses the electrical current from the EEC to position a piston within this sleeve, which opens or covers the port. In this way the current is related to flow by the shape of the port in the sleeve. A feedback device measures the position of the piston and the reading used by the EEC to assist in control, and to ensure that the position and hence flow control is operating correctly.

The servo supply is also used to power other hydraulic circuits within the unit, for example, the fuel shut-off valve. In response to electrical signals from the EEC, it can also power the actuators controlling, for example, variable stator vanes in the compressor.

Actuation

Actuators can use various power sources, in addition to high-pressure fuel:

> Pneumatic systems are simple and rugged, but heavy and relatively slow in response.

> Hydraulic systems offer high levels of power and response, at a low weight, but require complex ancillary equipment.

> Low-pressure fuel systems have relatively low power but are sufficient to move components such as inlet guide vanes.

> Direct electro-mechanical systems, which offered low response times and were relatively heavy, can now be replaced with more modern technology, ensuring that they are lighter, less bulky, and operate at significantly higher speeds.

Fuel pumps

The pumping system has to be able to deliver sufficient fuel flow to the engine under all conditions and at pressures high enough to overcome the gas pressure in the fuelspray nozzles generated by the engine compression system. Flow is also required to power the servo systems. The pumps are driven from the engine accessory gearbox.

The input pressure to the pumping system depends on the aircraft fuel system; in some cases, the engine pumps are required to continue to feed the engine with fuel if the aircraft fuel pumps fail – the engine pumps would then have to deliver fuel from the aircraft tanks. Generally, it is not possible to provide the suction and fuel pressure required with a single pump, and so two pumps are often used in series: a low-pressure centrifugal pump provides suction and delivers fuel to a high-pressure pump (most commonly a positive displacement gear pump), which in turn delivers fuel to the FMU.

In all cases, the pumping system must deliver enough fuel for the engine. The nature of the positive displacement pump sets a unique relationship between speed of rotation and flow delivered, resulting in more flow from the HP pump than is required by the engine. To accommodate this, the FMU returns unwanted fuel to the inlet of the HP pump, which then recirculates it. A consequence of this can be excessive heating of the fuel as it is repeatedly compressed and decompressed. A cooler is often required to dissipate this heat.

Bleed valves

Many engines include bleed valves in the compression system, which allow inter-compressor flows to be matched at low speeds. The position (open or closed) of these valves is signalled by the EEC, and fuel, electrical, or pneumatic servo systems are used to actuate them.

Electrical power supply

The reliability and availability of the electrical and electronic parts of the system are dependent on the quality and reliability of the electrical power supply. Each channel of the FADEC system requires independent power and, for aircraft, this is typically provided by a dedicated, gearbox-mounted generator. Aircraft power is also supplied for starting, and to provide a back up should the dedicated generator fail. In some military aircraft, only aircraft power is used, mainly because of the dependability of the aircraft supplies.

Software

The software embedded in the EEC defines the system behaviour. The performance of this software is therefore vital to the operation of the engine. The software is generated from the requirements using disciplined processes and extensive testing. These processes are defined in industry standard documents and guidelines. Software developed to these standards is expensive to generate and can take a considerable time, particularly due to the effort required in testing and qualification. Software tools and techniques are becoming available to reduce this effort but these are far from mature.

Indication systems

In modern systems, data from the control system and other sources is displayed on one or more display units mounted in the instrument panel; multi-function screens, which display basic engine data such as rotor speeds, turbine temperature, and power, have replaced the multitude of dials and individual instruments found in older aircraft. The multi-function screens are programmed to reconfigure themselves to display other data in response to abnormal circumstances, or as required by the operator. The information is displayed on the screen in the form of virtual dials with digital readouts and warnings; cautions and advisory messages are shown as text. A mimic diagram representing the physical layout of the equipment may be provided to assist in locating a problem. The displays are colour-coded and, when necessary, linked to audible warning systems so that the operator is aware of the severity of any problem.

In military aircraft, this data may be displayed using a 'head up display' (HUD). The HUD system projects information and instrument images onto the screen in front of the pilots.

Using this technology means that pilots do not have to divert their attention from the view around them. On some applications, information can also be shown on the visor as part of the headgear worn by the pilot.

Engine health monitoring

It is in the interests of all customers to minimise the cost of operation of the gas turbine and its associated equipment. The costs of operation include fuel, scheduled maintenance, and unforeseen events that result in the engine not being available when required. Monitoring systems can help to reduce all of these costs. Scheduling of major engine maintenance (for example, to restore performance after many hours of operation) is a complex economic decision for which monitoring systems can provide important supporting data.

Although it is not strictly part of the control system, the EHM electronics are often housed within the control system enclosure, and the two systems are to some extent integrated. It is important to note, however, that the safety requirements of the two systems are different and the design of each, and their integration, must reflect this. A function in the monitoring system cannot be adopted for use in the control system without considering the reliability of its implementation.

Data from the EHM systems are not generally available to the flight crew. Large amounts of data are stored although data reduction and analysis algorithms are used to make storage requirements more reasonable. Aircraft systems are used to transmit the data to a ground station, which in turn will forward the data to a centre where further analysis can be carried out in order to inform maintenance logistics.

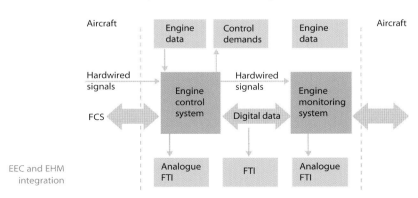

EEC and EHM integration

Temperature stations

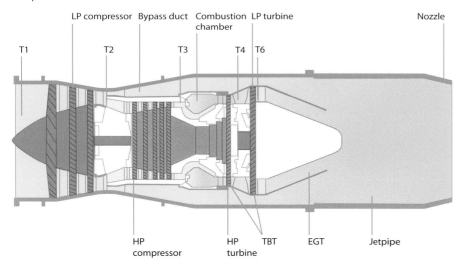

Probe locations

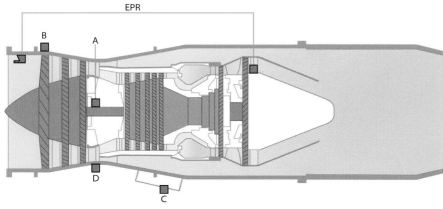

The advantages of these sensors are that they are resistant to damage (when in a housing) and give very accurate outputs with long-term stability. However, they have a slow response time when in a housing, need a constant current source to operate, and are relatively expensive.

Pyrometer
In this method, an optical device is used to view, for example, the turbine blades, connected to an infra-red (IR) detector by a fibre-optic cable. This method enables rapid, accurate measurement of temperatures. However, a compressed air supply is needed to keep the lens clean, and the output needs sophisticated signal processing.

Pressure sensors
Pressure sensors broadly divide into those required to provide high accuracy, and those that focus on transient response. Accurate measurement is required when pressure ratio is used to measure engine thrust. Transducers based on a variety of technologies are used for this purpose, but they generally need electronics to support their operation or to provide calibration information, and therefore the assembly is usually housed within the EEC, which can involve quite long pipe runs. If high bandwidth is required, simpler transducers, often based on strain gauge technology, are used and may be mounted close to the engine to avoid pipe delays.

Rotor speed sensors
Types of speed sensor include tachogenerators and magnetic variable reluctance (VR) probes.

A tachogenerator is a shaft-driven, electrical generator with a variable frequency output, which is related to speed. These devices are very rugged, but produce a relatively low output signal.

If a VR speed probe is used, it is positioned on the compressor casing in line with a small disc, which has accurately machined notches on its circumference and is mounted concentrically on the shaft. Rotation of the shaft results in a current being induced in the

Sensors
Whatever the particular application, a series of parameters needs to be measured at various locations around the engine system in order to control the engine and provide useful indication of performance to the operator. Typically, these are temperature, pressure, and speed measurements. The transducers used to take these measurements are chosen on accuracy, response time, and durability requirements.

Temperature sensors
Thermocouples
Thermocouples are used to measure high temperatures, typically at HP compressor exit, and in and around the turbines. The temperature of the gas in the turbine is measured at several radial and circumferential positions in order to even out any local temperature variations due to turbine entry temperature traverse effects.

Thermocouples have the advantage of being very reliable, small, and cheap; they also have a relatively quick response time over a large temperature range, and generate their own output – and so do not require an external power supply. However, thermocouples are easily damaged, and can lose accuracy through oxidation.

Resistance temperature devices
These devices are most often used to monitor engine intake air temperature. They consist of a platinum coil, exposed to the airflow, which changes its electrical resistance with temperature. Often, the device will consist of a single probe with a twin output and a heated body for anti-icing.

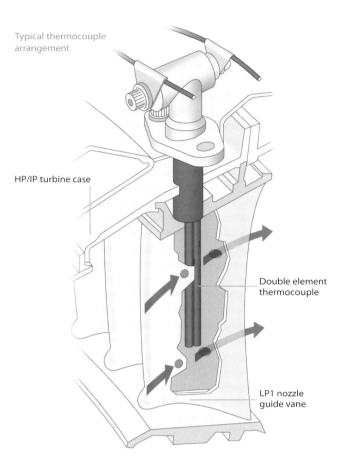

HP/IP turbine case

Double element
thermocouple

LP1 nozzle
guide vane

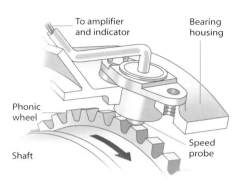

To amplifier
and indicator

Bearing
housing

Phonic
wheel

Shaft

Speed
probe

probe with a frequency content proportional
to engine speed.

Position measurement

Position measurement is used to confirm
that actuators are operating correctly and to
assist in closed loop control. There are three
main types of device used: the LVDT (linear
variable differential transformer), RVDT
(rotational variable differential transformer),
and the resolver.

An LVDT consists of three adjacent coils of
wire wound around a hollow form through
which a core of permeable material (such as
steel) can slide freely. The middle winding is
known as the primary coil, and is excited by a
relatively high frequency AC voltage. This sets
up a magnetic flux, which is then coupled
through the core to the other two, secondary,
windings, inducing a voltage in them. When
the moving core is centred between the
secondary coils, the voltage induced in them
is equal and opposite. If the core is displaced,
then an imbalance is set up, creating a
voltage that can be read and calibrated
to give a position.

RVDTs and resolvers are based on similar
principles, but are used to measure
rotational angles.

Vibration

Many engines are fitted with sensors that
continuously monitor the vibration level
of the engine. Indication of excessive
vibration is shown on the control display
unit using signals from engine-mounted
transducers. There are three main types
of vibration sensor:

> Piezoelectric accelerometers produce
a very low value charge signal through
deformation of a crystal lattice, and
require the vibration signal to be
processed using a charge amplifier
and sophisticated cabling.

> Piezoresitive accelerometers change their
resistance relative to an applied stress,
and are easy to use and install, but require
a separate power supply.

> Velocity pickups produce a voltage signal
from a magnet moving in a coil, are easy
to install, and require simple processing.

Safety and availability

Safety is the most important design
consideration in any gas turbine or installation;
another high priority is availability – the loss
of power from an engine, although not
necessarily a safety hazard, can cause severe
operational disruption. The duplication of the
electrical elements of the system is evidence
of this concern. Rigorous analyses and testing
are necessary to ensure that faults in the
system are correctly accommodated to allow
for continued engine operation.

Just as it is safe to complete a flight during
which a failure has occurred in the duplicated
part of the system, it can also be shown by
analysis that the aircraft can continue to
operate for subsequent flights for a defined
period before a fault is repaired. A 'time
limited despatch' analysis is carried out to
establish which faults can be treated in this
way and for how long. This is of considerable
benefit to the aircraft operator, who can
continue to operate the aircraft normally
and repair the fault at a convenient time,
for example, when the aircraft next returns
to the operator's main base.

Other safety features may also be required,
implemented either in the software or in
dedicated hardware to address the effects
of adverse operating conditions, or of
particular engine or control system failures,
which could represent a threat to the aircraft
if not accommodated.

Defence applications

Much of the control system technology used in military applications of gas turbines is similar to civil aerospace engines. However, engine requirements can differ markedly between different military applications, depending on whether the aircraft is a single-engine trainer, a large, twin-engine fighter with full afterburner capability, or a propeller-driven military transport.

Afterburning (>> 243), also known as wet thrust or reheat, requires additional fuel handling equipment such as pumps and metering valves. They employ similar technology to that described above. Afterburning also requires a variable area exhaust nozzle in order to control the LP system working line. In some applications, the final nozzle is not only variable in area, but the thrust can be vectored by limited angular displacement of the nozzle. This enables a significant increase in aircraft agility without the use of large control surfaces and their associated drag. A variable area nozzle is controlled with actuator rams, typically powered by HP fuel pressure, and an appropriate servo system signalled from the EEC.

The application of the aircraft may also involve vertical or short take-off, hover, and vertical landing. There are a number of aircraft configurations used to deliver this functionality. Each requires different levels of power to be extracted from the engine to provide vertical thrust, and to stabilise and manoeuvre the aircraft until sufficient forward speed is attained that the conventional flight control surfaces become operational. The control system must ensure the engine remains stable during these manoeuvres and can respond to the very rapid changes in power that are required.

A single-engine aircraft often warrants additional system provisions. This might take the form of a mechanical system which can be invoked by the pilot should all electronic means of control fail. However, as the functions required of such a controller increase, for example, in a complex STOVL application, a mechanical solution is not possible, and so the safety case must be justified based on the electronic system's reliability and built-in redundancy.

Helicopter systems

In many respects, helicopter control systems function in much the same way as those of fixed-wing aircraft – sensors monitor engine parameters, which are communicated back to an engine controller. However, the nature of a helicopter and its engine configuration mean that there are different control system requirements.

The engine controller must closely control the engine in order to provide a stable power turbine shaft speed. This then allows a constant helicopter rotor speed, while the pitch of the rotor blades controls lift and horizontal helicopter speed.

Traditional control systems operated with a throttle, using the collective pitch lever as the main load demand, with a twist grip for the pilot to trim the demand and keep the rotor speed within defined limits. Modern engines do not have a conventional throttle; they operate on a governing system whereby the pilot demands a load and the control system and inherent control laws and schedules will control the engines to maintain the correct rotor speed. In such a system, the power turbine speed and torque are monitored and the fuel flow is modulated accordingly.

One of the key aspects of helicopter engine control is matching the torque provided by the engines on multi-engine aircraft. Torque mismatches can provide significant aircraft performance penalties. Engine parameter matching, through communication of data between the engines, can be used in order to enable isochronous control, and maintain an even loading of torque.

Although vibration absorbers can be used in some cases, helicopters experience significant levels of vibration; the engines are mechanically coupled through the drive train to the rotors and consequently there is very little vibration damping. It is therefore important to monitor vibration levels.

Marine systems

The environment around a marine gas turbine is somewhat more benign than most aircraft applications; with fewer constraints on weight and space, there is more scope for installation options and therefore the use of less rugged equipment. Even so, it is common for these products to use existing aerospace components, particularly in the fuel system. These may be mounted on the engine or assembled onto a fuel 'skid' mounted in the engine enclosure. Similarly, electronics can be engine-mounted within the enclosure or outside in conventional equipment racks, or may be split between a number of these locations, communicating by means of digital data busses.

The marine gas turbine may mechanically drive a propeller, water jet, or other propulsion system, or may drive an electric generator providing power to electric motors, which in turn provide the motive power. The engine controller is therefore required to interface with the control system of this additional equipment in order to optimise overall performance. In addition, there is usually a separate system providing the human-machine interface – the means by which the ship's crew provide inputs to the propulsion system and monitor its performance. The system may need to be operated from several different locations on the ship: the engine room, the captain's chair in the centre of the bridge, or from bridge 'wings' during close harbour manoeuvring.

Energy systems

Fixed-installation gas turbines, used for pumping large volumes of gas or fluid fuel, or for electricity generation, have many similarities to marine installations. One key difference, however, is that fixed installations are subject to much more stringent emissions regulations. This leads to complex combustion systems and consequently complex control requirements.

In some installations, an engine may be required to operate on a variety of fuels from gas to diesel oil. This adds considerable complexity to the pumping, metering, and piping arrangements for the fuels: when an engine changes fuel, the systems have to be primed, change-over achieved, and then the system for the now unused fuel purged for safety reasons. This all adds complexity to the control system requirements.

The gas turbine will be part of a package designed to deliver power in the form required by the customer. The control of the package can be comparable with the complexity of the gas turbine control and the whole assembly can only be effective if these two systems are designed to work together.

203

This completes component definition, production is a new challenge.

manufacture and assembly

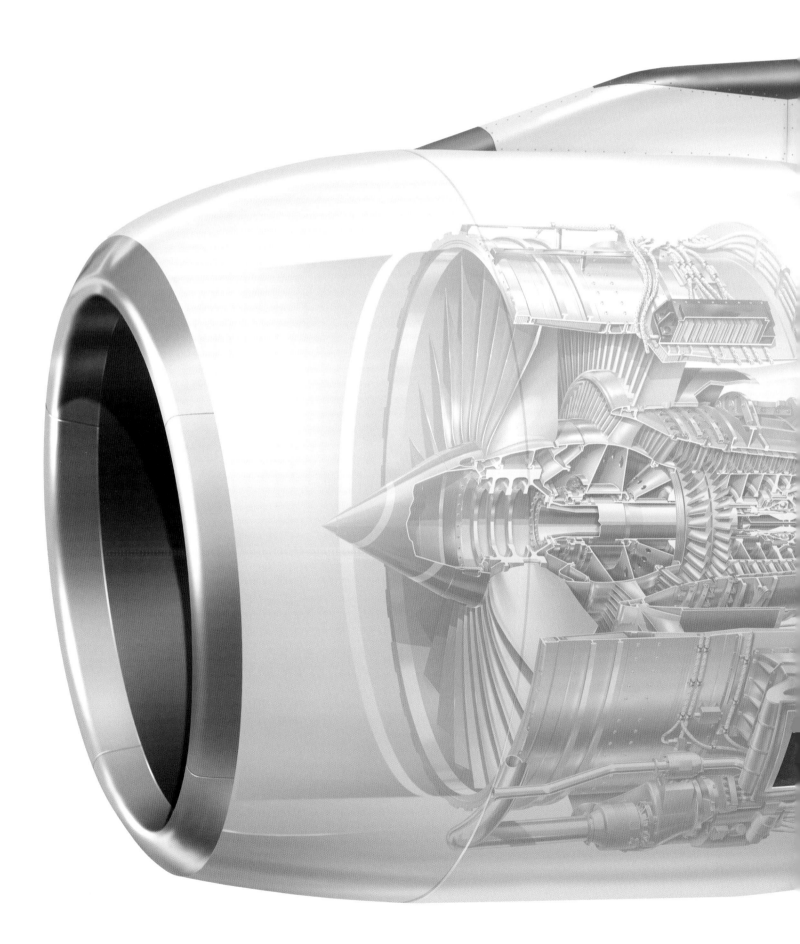

Delivering customer benefits **in service**
demands vision, versatility, and reliability.

IT MIGHT BE POSSIBLE TO DESIGN A THEORETICALLY PERFECT ENGINE: IT WOULD NOT NOW BE POSSIBLE TO MAKE IT – AND, IN ALL PROBABILITY, NEVER WILL. THE CHALLENGE OF MANUFACTURING IS TO PRODUCE, IN A PREDICTABLE AND REPEATABLE MANNER, AN ENGINE AS NEAR AS POSSIBLE TO THE ENGINEERING IDEAL.

manufacture and assembly

Gas turbine manufacture is a global enterprise; this globalisation has been enabled and promoted by the advent of rapid secure electronic communication and the standardisation of data formats.

Throughout the design and development stages of a gas turbine, close liaison is maintained between

> design
> manufacturing
> development
> product support
> the supply chain
> the customer

to ensure that the final design satisfies the engineering specification, manufacturing process capability, delivery, and cost targets.

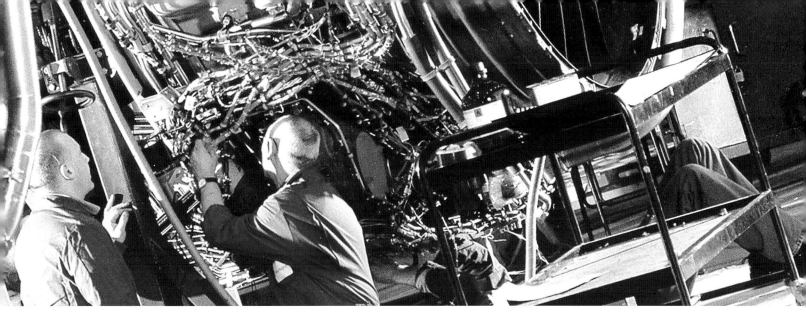

Each component is manufactured to provide the highest possible performance and mechanical integrity through a long service life at the lowest possible cost and weight. Consequently, the methods used during manufacture are diverse – usually determined by the characteristics of each component such as shape, surface finish, geometric tolerance, and material properties.

No manufacturing technique or process that offers any sort of advantage is ignored. Continuous improvement is a given, and considerable resource is invested in developing and implementing innovative manufacturing technology.

Materials

Engine materials are chosen primarily for their ability to withstand the environment in which they are required to operate. Consequently, strength at temperature and corrosion resistance are major considerations. Unfortunately, materials with properties that make good engine components often present a manufacturing challenge.

In order to minimise costs, it is important to acquire material as close as possible to the net shape of the component, not only to minimise material procurement costs but also to minimise subsequent process costs such as machining, inspection, and heat treatment.

Unfortunately, there is a gap in the manufacturing ability to create some components with the desired mechanical properties without some waste. A trade-off between material properties and ease of manufacture is an ever-present fact of life.

Improvements and step changes are constantly and actively pursued, although the method of manufacture for most families of gas turbine components is now well established.

Combustor and turbine casings are made from either ring forgings, or fabrications, or a hybrid of the two. Some compressor casings are cast; discs and shafts are machined from very high quality forgings.

Cold components are made mainly from titanium alloys; less often from aluminum and magnesium alloys. Increasingly, composite materials are finding application in this area as higher temperature composite materials become commercially available.

Ceramics are also becoming more common, particularly in the form of temperature-resistant coatings, wear-resistant surfaces, and lightweight, rolling elements in ball bearings.

Material properties dictate that hot engine components are produced mainly from nickel and cobalt alloys; some temperature-resistant steels are still in use, for example, for bearing tracks, shafts, and discs.

Combustion and HP turbine components operate in high gas temperatures relative to

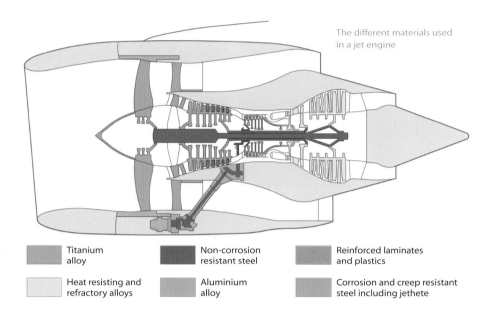

The different materials used in a jet engine

▮ Titanium alloy	▮ Non-corrosion resistant steel	▮ Reinforced laminates and plastics
▮ Heat resisting and refractory alloys	▮ Aluminium alloy	▮ Corrosion and creep resistant steel including jethete

their melting point and need to be engineered to perform and survive in this environment. Typically, HP turbine blades and nozzle guide vanes (NGVs) have complex internal cooling passages and utilise surface boundary film and effusion cooling as well as ceramic and intermetallic coatings for heat and oxidation resistance. Such components also have single crystal or directionally solidified structures to maximise their strength. Casting is the only way to manufacture such structures.

Casting

Casting is one of the oldest metal forming processes known to man. The process has evolved to produce components with high standards of surface finish, complex internal passages, repeatable accuracy, amazing surface detail – and it is still being developed.

In engine manufacture, castings can be divided into two families: structural castings and 'hot end' components. Both use investment casting technology where a

highly accurate, hollow, ceramic mould or shell is created and filled with molten metal to create a component. The shell not only contains the surface detail but can also include internal details created by the incorporation of complex and delicate cores. Core technology has been a key element in enabling the manufacture of highly sophisticated cooling systems. The main requirements are, first, the ability to position the core within the mould and, second, for the core to maintain its shape and position during the mould firing, filling, and metal solidification phases of the operation.

Right: A ceramic shell core that gives the complex internal cooling geometry to an HP turbine blade

Far right: A ceramic shell for four HP turbine blades ready to go to casting. The ceramic cores are already inside each shell.

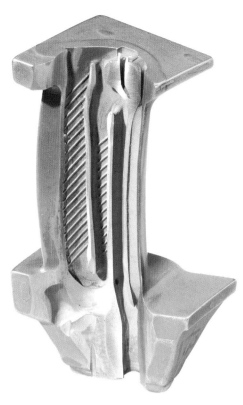

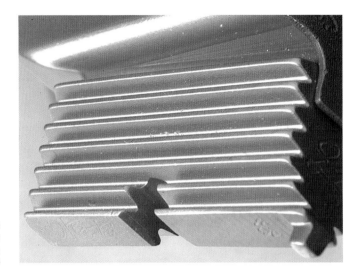

Left: A section through an HP turbine blade showing the complex cooling geometry after casting

Right: A fir-tree blade root of an HP turbine blade finished using a grinding process

Structural castings are usually complex equiaxed castings with coaxial, annular features, such as the intermediate compressor casing, or compressor casings with numerous variable vane spigots or integral outlet guide vanes. Combustion headers and chambers also fall into this category. For industrial applications, casings for power turbines and barrel compression units are generally large structural castings.

Hot end castings comprise a range of components such as combustor tiles, NGVs, turbine seal segments, and turbine blades; they are usually cast in a vacuum to prevent oxidation.

To extend cyclic life, turbine blades and NGVs are cast so that they either contain no grain boundaries (having been formed from a single crystal) or contain crystals orientated in a pre-determined manner by the cooling method (directionally solidified). The moulds used for directionally solidified and single crystal castings differ from conventional moulds in that they are open at both ends; the base of a mould forms a pocketed bayonet fitting, into which a chill plate is located during casting.

Metal is introduced from the central sprue into the mould cavities via a ceramic filter.

These – and any orientated seed crystals that are required – are assembled with the patterns prior to ceramic coating. Extensive automation ensures the patterns are coated consistently with the shell material.

Developments in rapid prototyping have led to the use of stereo-lithography in the manufacture of moulds. A computer-controlled laser is used to selectively solidify UV sensitive resins, creating 3D shapes and so enabling moulds to be made from ceramic-filled resins without the need of wax patterns. This removes a significant number of operations from the traditional investment casting process.

Machining

To achieve the precision fits demanded by the jet engine, some form of machining has to be undertaken on all components. Availability of high-speed, multi-axis, computer-controlled machine tools, using ceramic and intermetallic cutter materials with high-pressure coolants, has resulted in chip machining competing successfully with processes such as chemical and electrochemical machining, which, historically, were used primarily because materials were too tough to machine by more conventional processes. Chip machining is now used, for example, to remove metal around casing bosses and to machine holes for casing isogrid patterns.

Fixturing

The drive to lean manufacture and minimum inventory holding has increased the demands on component fixturing and setting times.

Through the integration of coordinate measuring machines (CMMs) and computer numerical control (CNC) machines with robust, individually identified fixtures, it is now commonplace to create a unique machining program for each component so optimising the position of a feature relative to a datum.

An example is the grinding of a blade location fixing relative to its aerofoil. Typically, aerofoils are finish cast or forged and their fixing needs to be machined to ensure the aerofoils will locate in the correct position and attitude in the engine. Loading a component into a fixture, determining its position relative to the fixture datum, and then adjusting the machining programme to accommodate variations, achieves a rapid throughput of parts with a high conformance rate and minimal operator input or intervention.

Grinding

Developments in grinding technology such as continuous-dress, creep-feed grinding have revolutionised the metal removal rate and machining capabilities of the grinding process.

Cast turbine alloys are particularly difficult to machine but grinding using open structure, vitreous bonded wheels and computer-directed, high-pressure coolant on purpose-built machining centres has enabled these components to be produced rapidly in very few operations. CBN (cubic boron nitride) plated grinding wheels can also be used to produce accurate features in jet engine components.

Drilling cooling holes

The integrity of the engine relies heavily on controlling the temperature of components. Active cooling is achieved by passing cooler compressor air through hot components; however, this air is lost from the overall engine cycle and consequently must be minimised. To do this, large numbers of small holes are preferred to cool the maximum volume with the minimum amount of air. Typically, tens of thousands of cooling holes are required within the combustion and turbine components.

Historically, non-conventional drilling techniques such as electro discharge machining (EDM) and electro chemical machining (ECM) have been adopted as the only viable drilling processes available. Lasers have now joined this list. The key issues with cooling hole drilling are hole integrity and avoidance of damage to internal passages in order to maintain airflow within strictly controlled limits.

EDM and laser drilling are both thermal processes, which melt and volatilise material, producing a heat-affected zone and a recast layer. Acceptance standards have been established for these effects and dictate which process can be applied. ECM dissolves material electrolytically, so there are no thermal effects and little or no wear to the tool. EDM removes metal from the workpiece by converting the kinetic energy of electric sparks into heat as the sparks strike the workpiece. Sparks will occur when a sufficient build-up of electrons has enough energy to jump across a gap where there is an electric potential between two conducting surfaces: the electrode and the workpiece.

Electrons break through the dielectric medium between the conducting surfaces and, moving from negative (the tool electrode) to positive (the workpiece), strike the latter surface with great energy. The amount of work that can be effected in the system is a function of the energy of the individual sparks and the frequency at which they occur. Because of the heat generated, EDM electrodes wear and are treated as consumables.

A schematic of a high performance grinding system

Open structured wheel having strong abrasive retention

High pressure coolant jetted into grinding wheel

Coolant nozzle positioned close to wheel and point-of-cut

Coolant remains through duration of cut

Coolant forced out of grinding wheel by gravitational force

Depth of cut

Cooling holes drilled into HP turbine blade

EDM drilling electrodes can be solid wire rods or hollow tubes. Both EDM and ECM impart the electrode shape to the workpiece. To produce a hole, the electrodes must be fed into the workpiece as material is removed in front of the electrode. When drilling by EDM and ECM, it is usual to use multiple electrodes to maximise drilling rates while guides are used to control the position and direction of each electrode.

By using both multi-channel power supplies that manage the power supplied to each individual EDM electrode and also hollow electrodes with a high dielectric pressure to aid flushing, very high drilling rates are achieved. Because electrical activity is monitored in each electrode, the point of breakthrough can be detected, and the operation terminated before any damage occurs to the far wall. The most recent EDM drilling machines use environmentally friendly, deionised water as the dielectric fluid rather than the paraffin or silicone oils used in earlier machines.

Unlike EDM and ECM, a laser does not require the workpiece to be electrically conductive; therefore, the drilling of non-metallic materials becomes possible and the process is used for drilling ceramic-coated components such as combustion chambers. A laser is also a single-point tool and so requires less investment in component-specific tooling. However, to compete in terms of holes per minute, it must be able to drill rapidly.

Laser drilling in action

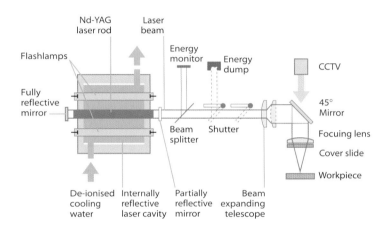

A laser drilling system

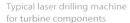

Typical laser drilling machine for turbine components

Although there are many laser sources available, most laser drilling machines use a pulsed, solid-state laser in which the lasing medium is neodymium in the form of a man-made, neodymium-doped, yttrium aluminium garnet rod. This typically emits light with a wavelength of 1064 nanometres, so is infra-red and invisible to the human eye. When pulsed with high-powered flash lamps and focused to a point, the Nd-YAG laser produces a pulse, of energy that will vaporise most materials instantaneously.

It is essential to understand how well the laser beam couples with the target material. This is a function of angle of incidence, surface texture, and wavelength. Short wavelengths, unpolished surfaces, and a 90 degree angle of incidence give optimal results.

Laser hole drilling can be achieved by one of two methods: percussion drilling or trepanning. Percussion drilling, as the name suggests, entails hitting the workpiece with the laser beam to create a hole. Trepanning creates a small hole, and then generates the desired hole size using a rotary motion.

Percussion drilling is fast, but produces a tapered hole with a thicker recast layer. Trepanning produces a better hole shape, but is slower. Laser beams are difficult to arrest after the drilling process is completed; damage to material directly behind the section being drilled can, therefore, be a problem – although materials such as PTFE (Polytetrafluoroethylene) are good at absorbing laser energy and are used where access allows. Laser technology is still evolving rapidly and developments such as pulse shaping, twin rod, diode-pumped, and frequency-doubled lasers are at different stages of implementation. All offer improvements in drilling rate and efficiency.

The plasma welding process

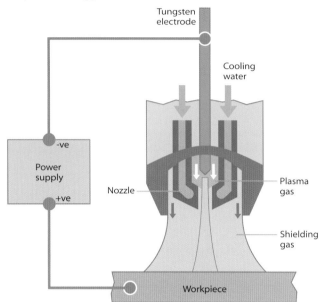

Welding with a plasma torch[*5]

Joining

Cooling holes are not the only type of hole found in a jet engine. In order to facilitate assembly and maintenance, hundreds of bolt holes are required. These holes tend to be conventionally produced by drilling and milling. Mechanical fasteners, however, add weight and require space, so where possible joining techniques such as welding, bonding, and brazing are used.

Tungsten inert gas (TIG) welding is the most common form of fusion welding in use and is the most economical method of producing high-quality welds for the range of high-strength, high-temperature materials used in gas turbine engines. For this type of work, high-purity argon, shielding gas is fed to both sides of the weld and the welding torch nozzle is fitted with a gas lens to ensure maximum efficiency for shielding gas coverage. A consumable, four per cent thoriated tungsten (addition of thorium oxide to the tungsten) electrode, together with a suitable non-contact method of arc starting is used. To prevent the formation of finishing cracks, the weld current is reduced in a controlled manner at the end of each weld. Whenever possible, a combination of mechanised welding with a pulsed arc is preferred. TIG welding is used on sections up to three millimeters; for thicker sections, plasma welding can be used. Plasma welding

is an electric arc process similar to TIG except that the current is carried by the plasma generated within the torch.

Electron beam welding (EBW) is used to join thicker sections with high-quality welds, minimal distortion, and a reduced heat-affected zone. The process uses a high-power density beam of electrons to join a wide range of different materials of varying thickness. The welding machine comprises an electron gun, optical viewing system, work chamber and handling equipment, vacuum pumping system, high or low voltage power supply, and operating controls.

Major rotating assemblies for gas turbine engines such as intermediate- and high-pressure compressor drums are manufactured as single items in steel, titanium, and nickel alloys and joined together by EBW. This technique allows design flexibility as distortion and shrinkage are reduced and dissimilar materials, serving quite different functions, can be homogeneously joined together. For example, HP turbine stub shafts requiring a stable bearing steel can be welded to a material that can expand with the mating turbine disc.

Computer numerical control (CNC) for work handling, seam tracking to ensure the joint is

accurately followed, and closed loop control of the under-bead part of the weld, guarantee that the full depth of material thickness can be welded accurately in a repeatable process. Careful design of joint geometry, coupled with fixtures that are capable of being remotely manipulated within the EBW chamber, enable a series of joints to be completed with the minimum number of operations.

The electron beam welding process

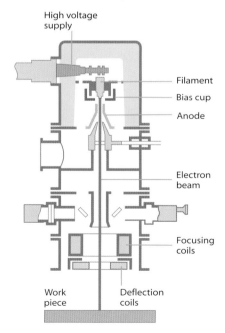

A typical electron beam welding machine

Two discs in a compressor drum joined together using electron beam welding

An example of additive manufacture by TIG weld deposition: an extended intercase, with bosses, prior to machining

Section through a hollow fan blade

A hollow wide-chord fan blade

The emergence of high-powered lasers, particularly continuous wave, solid state lasers, could provide a lower cost alternative to EBW.

All welds are visually and penetrant inspected. In addition, welds within rotating parts, such as compressors and turbines, and welds within pressure vessels are radiologically examined.

Fabrication using joining processes has long been recognised as an efficient way of utilising raw materials. The overall strategy is to put metal where it is required; however, fabrication invariably means manufacturing sub-assemblies, which may need trimming and machining.

Developments in computer simulation and automation has enabled deposition of metal directly in three dimensions so generating components with little or no fixturing close to their finished shape. Components can be built up from scratch onto a base plate or features such as flanges and bosses can be added selectively onto pre-existing components. In both cases, material is deposited continuously in layers until the final shape is created.

Additive manufacture can be achieved either by using welding processes such as wire-fed TIG, MIG (metal inert gas), and more recently by powder-fed laser fusion. The use of wire or powder means that a variety of components can be made from a common stock of raw material.

EBW, TIG, plasma, and laser welding are all examples of fusion welding that involve melting and re-solidification of the materials being joined. In contrast, solid-state bonding processes such as inertia, friction, and diffusion bonding rely on atomic migration across the joint interface and will produce joints in alloy combinations that fail to fusion weld.

The key requirements for solid state bonding are intimate contact, surface cleanliness, and atomic diffusion. Intimate contact is achieved by ensuring good fits and the application of pressure. Surface cleanliness is achieved by chemical cleaning or the expulsion of oxidised material by extrusion. Atomic diffusion can be initiated by heat, mechanical work, or a metallic chemical activator.

Diffusion bonding is used in the manufacture of hollow titanium fan blades and outlet guide vanes. The process allows two or more sheets of titanium to be joined in chosen areas to form a monolithic structure that when cambered, twisted, and superplastically blown forms a hollow wide-chord fan blade.

Lightweight structures can also be created by the use of honeycomb sandwiches. Typically, these brazed or activated diffusion bonded structures are used in areas where high stiffness and minimum weight is required. To achieve intimate contact during brazing and diffusion bonding of large areas, gas pressure is applied either by use of a pressurised furnace or gasbags.

Fan blisk machined from solid
on a 5-axis milling machine

Inertia and friction bonding use higher
forging loads to achieve high-integrity bonds
in a rapid, phased sequence of events. Initially,
the joint faces are brought into contact with
a moderate load; relative motion commences,
and heat is generated due to friction. This heat
softens the interface material, which extrudes
as 'flash'. In the final phase, the bond is created
when a high forging load is applied and
relative motion ceases. Flash creation
means that some material loss must be
accommodated, but it ensures intimate
contact is maintained and any contaminants
at the joint interface are expelled. The cycle
takes seconds to complete; the join has
a very fine grained structure and a narrow
heat-affected zone. The speed and integrity
of this process lends itself to use on critical
parts such as disc to shaft, disc to disc, and
blade to disc joints. On round components,
rotary motion is used and is relatively easy
to control. On rectilinear joints, such as those
where blades are bonded to a disc to make
an integrally bladed disc or blisk, a more
complex linear motion is used.

Above: Blisks manufactured
using linear friction welding

Blisks

Blisks are far lighter than equivalent
conventional bladed discs because removing
the need for mechanical fixings means that
hub diameters can be significantly reduced.
By integrating technologies, hollow-bladed
blisks can be manufactured.

To accommodate the bonding process, additional material at the foot of the blade is necessary and subsequently has to be machined away to produce an aerodynamic blend between the aerofoils and disc rim. Smaller blisks tend to be machined from solid. For both solid and bonded blisks, complex 5-axis milling is required to generate the finished shape. Tool path programming, collision avoidance, and on-machine measurement all facilitate blending, which is critical to the successful manufacture of blisks. Programming also controls the quality of surface finish.

Surface finish

Surface finish affects the aerodynamics of an aerofoil and most aerofoils undergo a finishing treatment such as barrelling, vibro-polishing, shot peening, or vapour blasting to produce a uniformly smooth surface. Blasting also imparts compressive stress, which is beneficial to fatigue life.

Laser shock peening can impart very high levels of compressive stress and is used in areas that are sensitive to fatigue or crack propagation. The process converts the energy in a pulse of laser light into a shock wave by using a film of water to direct the explosion that occurs when the laser strikes an ablative, or sacrificial, medium on the surface of the component. To impart the necessary energy, the laser is focused onto a spot, and in order to peen an area, a pattern of overlapping spots is applied.

Composite materials

High power-to-weight ratios and low component costs are very important considerations in the design of any aero engine, particularly when the engine is used to power V/STOL aircraft where weight is critical. Composite materials allow the designer to produce lightweight structures in which strength in any direction can be varied by the directional lay-up of fibres according to the applied loads. Composite materials have replaced and will continue to replace steels and titanium in a variety of aerospace components, including bushes, unison rings, and bypass duct assemblies.

Conventionally cast and fabricated casings and cowlings are also being replaced by casings of a sandwich construction that provide strength and lightness, and which also act as a noise suppression medium. These casings comprise a honeycomb structure of aluminium or stainless steel interposed between layers of dissimilar material.

Inspection

To ensure conforming components and assemblies are produced, all parts need to be inspected for both dimensional accuracy and flaws such as cracks and internal defects. Non-conformance also impacts on cost and productive capacity. Consistency of manufacture can be statistically determined and a trend established that can identify when pre-emptive remedial action is necessary before an acceptance threshold is crossed.

Dimensional conformance is assessed by a wide range of methods. High-volume parts are best suited to the use of automated loading and dedicated gauges with multi-directional probes that are able to measure a number of dimensions simultaneously. On lower volume parts, automated inspection is applied either in the form of a coordinate measuring machine or by using CNC machine tools capable of using a measuring probe as part of their tooling suite.

CMMs primarily use touch probes however, non-contact techniques such as triangulation and photogrammetry are becoming more common.

Component integrity is assessed by non-destructive techniques such as ultrasonics, radiology, magnetic particle, eddy current, and penetrant inspection, as well as electrolytic and acid etching. Computer x-ray topography, real time x-ray, and thermography are developments that are making non-destructive testing faster, better, and cheaper.

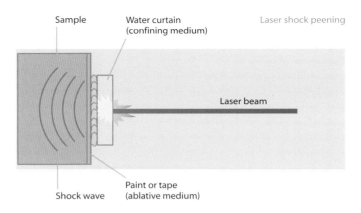

Laser shock peening

Sample — Water curtain (confining medium) — Laser beam — Paint or tape (ablative medium) — Shock wave

The modern factory

Final assembly of
a two-shaft V2500

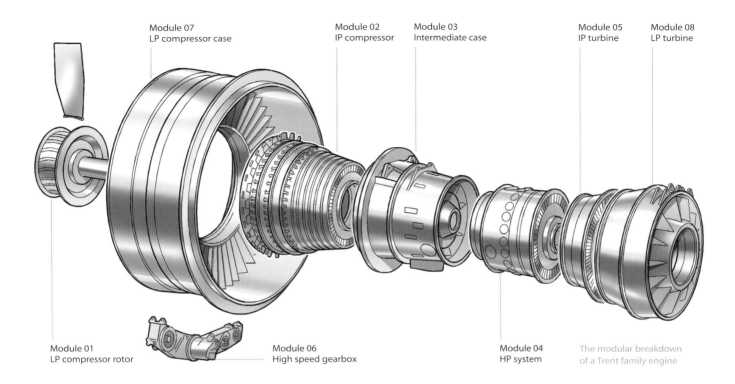

Module 07
LP compressor case

Module 02
IP compressor

Module 03
Intermediate case

Module 05
IP turbine

Module 08
LP turbine

Module 01
LP compressor rotor

Module 06
High speed gearbox

Module 04
HP system

The modular breakdown
of a Trent family engine

Assembly

Once the thousands of individual components have been manufactured, the complete engine is assembled. Most modern engines are designed as a series of modules for ease of assembly and subsequent maintenance. This approach naturally leads to two distinct assembly processes: module assembly and engine assembly.

Module assembly

On the Trent family of engines, the modules are split as follows:

> Module 01: LP compressor

> Module 02: IP compressor

> Module 03: Intermediate case (intercase)

> Module 04: HP system

> Module 05: IP turbine

> Module 06: High speed gearbox

> Module 08: LP turbine.

Module 01: the fan consists of a number of fan blades, fillers, the fan disc, and the front and rear plates. These are assembled and then connected to a polar moment of inertia

(PMI) machine, which simulates the LP shaft for balance purposes. The assembly is then taken through a process that removes the out-of-balance effect of the fan assembly so that it is within the limits defined by the design requirements.

The LP shaft is assembled and put through a similar process using a PMI machine that represents the fan, with the intent of removing the out-of-balance in the LP shaft assembly.

Module 02: the IP compressor consists of the IP compressor rotor, IP compressor case, and front bearing housing. The IP compressor rotor assembly process includes reaming the curvic coupling to the rotor drum, balancing at various stages of build, and blade tip grinding – a process that reduces the length of the rotor blades to a predetermined standard size. The final assembly operation for the rotor is, as with all rotating assemblies, the removal of the out-of-balance.

The IP case consists of three separate cases: the front bearing case – with one stage of

A Trent 800 IP compressor drum showing IP1 and IP6 stages without blades mounted.

OGV assembly

Fan blade and disc assembly

variable-vanes, the front case – with two variable stages, and the rear case with six stator stages. The case is assembled and, as with the IP rotor, all the rotor blade paths are machined to a predetermined standard size to match the IP rotor blades when assembled.

Machining of rotors and cases, once assembled, enables the removal of any build-up of component tolerances, and the assemblies to be machined to the optimum size for compressor efficiency; as the sizes are standard, mini modules can be interchanged.

The front bearing housing holds the front bearings for the LP compressor and the IP compressor. This assembly also contains the variable inlet guide vanes (VIGVs) and the shaft speed sensors for the LP and IP shaft.

Once all three of the main assemblies are completed, they are assembled as a complete module, with the associated control mechanisms for the three variable vane stages.

Module 03: the intercase is in the centre of the engine and holds the main thrust bearings for all three rotating systems. The three main assemblies are the LP/IP shaft assembly, the HP shaft, and the case assembly. The case assembly contains an internal gearbox to allow drive to be taken from the rotating shafts to drive the high-speed gearbox (module 06).

The shafts are assembled using dimensions from the case to ensure that the axial positions of all three rotating systems are set. To ensure that all the components are correctly seated, they are assembled using a hydraulic press. After the assembly of the shafts, where required, the residual unbalance is removed.

Module 04: this consists of the HP compressor rotor, HP compressor case, combustion system, HP NGVs, and the HP turbine rotor.

The HP rotor and cases are assembled in

the same way as the IP rotor and case. The HP case links the intercase module (03) to the IP turbine module (05).

The combustion system consists of an inner combustion case, which contains the outlet guide vanes from the HP compressor, an outer combustion case, and the annular combustion chamber containing the fuel spray nozzles and igniters. These are assembled during the module final assembly.

The HP NGV assembly contains the inner rear combustion case and the HP NGVs. The HP turbine rotor consists of air-cooled blades attached to a disc that is connected forward to the compressor mini-disc and rearward to a stub shaft that is located by a roller bearing in the HP/IP hub.

Module assembly starts with the assembly of the HP compressor rotor and cases, then the combustion system assembly is added, followed by the HP NGV assembly; finally, the HP turbine is fitted. The whole assembly is then checked that it has been built straight and concentrically.

Module 05: the IP turbine consists of a casing, blades, IP NGVs, the first stage of LP NGVs, the turbine disc, a shaft connecting to the IP compressor stub shaft via the thrust bearings in the intercase, and the rear roller bearings for the HP and IP shafts.

The IP turbine disc and shaft are connected together, the blades assembled, and the out-of-balance removed. The IP turbine case consists of an HP/IP bearing support, IP NGVs, and the outer IP turbine case. The bearing pack is assembled to the HP/IP support, the NGVs are then fitted, and the whole assembly is fitted into the outer case. Finally, the IP turbine is fitted into the IP case assembly.

Module 06: gearboxes are generally manufactured by specialist gearbox manufacturers and delivered as complete certified modules. It is mounted on the lower part of the LP compressor case and is driven from the internal gearbox in the intercase (03) via the radial and angled drive shafts.

Module 08: this consists of LP turbine rotors, LP NGVs mounted to the LP turbine case, and the rear bearing support assembly. The exact number of rotors and NGV stages varies with the engine type. This assembly contains the tail bearing housing which supports the rear LP roller bearing.

The LP turbine assembly starts with the assembly and balance of the individual turbine discs as they are put together. The next stage is that of assembling the complete turbine discs into the LP turbine case (NGVs mounted already) and fitting the LP turbine shaft, before final removal of any residual out of balance.

The rear bearing support assembly is assembled in the same way as the IP turbine case and vanes, except that instead of NGVs, there are sheet metal fairings that protect the bearing support struts and direct the gas flow. Module 08 operations finish with fixing of the rear bearing support assembly to the LP turbine case.

Engine build
Engine build has three main elements:

> core assembly

> module 07 and LP compressor case assembly

> final engine assembly.

Core assembly is the assembly of the core modules, described above, in the following order: module 02 is fixed to module 03 and then module 01 (LP shaft only) is fitted. The assembly is rotated so that the module 03 is uppermost and the modules 04, 05, and 08 are assembled sequentially. The core of the engine is then dressed with connecting pipes and harnesses before being placed in flight position ready for the connection of the fan case.

Final assembly starts with the fan case being connected to the core

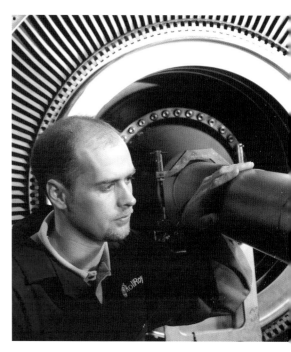

Module 08 – a Trent 800 LP turbine assembly.

Module 07: is the largest module and is an assembly of front and rear casings and the fan outlet guide vanes (OGVs); it is usually referred to as the fan case. The front casing must contain a fan blade released during engine running; more prosaically, its constituents include acoustic panels to minimise noise emission, ice impact panels, and the fan track lining to reduce tip losses. The rear casing carries the fan case-mounted accessories.

The LP compressor case, is assembled in parallel to the core build. The assembly consists of the fan case, module 06, and the external accessories with connecting pipework and harnesses. These are assembled in this order and the assembly is placed in flight position ready to connect to the core.

Engine final assembly starts with the connection of the fan case to the core. Then the final engine dressing is completed with the remaining pipes and harnesses. After this, module 01 (fan assembly) is fitted.

Having completed the engine assembly, the engine is then prepared for pass-off testing by attaching it to an engine pylon that simulates the conditions of its destined airframe. Once the engine has been through pass-off testing, it is ready for dispatch to the airframer or airline to enter into service.

The gas turbine is now complete – and useless until installed where it can be useful.

installations

THE JET ENGINE IS NOTHING IF NOT VERSATILE. IT CAN BE DEPLOYED ON OIL AND GAS PLATFORMS, IN POWER STATIONS AND SHIPS; IN THE AIR, IT CAN PROVIDE FORWARD, VERTICAL, VECTORED, AND REVERSE THRUST. THIS VERSATILITY PRESENTS A VARIETY OF INSTALLATION CHALLENGES.

installations

Nacelles and fuselage intakes

In most civil installations, the engine is enclosed within a nacelle. This is mounted on a pylon from either wing or fuselage, and supplied with air via a pitot intake. In military installations, which tend to have higher flight speed requirements, the engine is normally enclosed within the fuselage or wing root; therefore, air must be supplied to the engine via a more integral intake.

The Trent 800 short nacelle on the Boeing 777 gives separate jets for core and bypass flows

Vertical/short take-off and landing application in the Harrier

Installations

All installations require interfaces between the engine and the application, and protection of the engine from hazards such as fire and icing. The installation also has to ensure that the engine is fully integrated with the application, fulfilling its design requirements: weight and aerodynamics are key considerations on an aircraft; energy and marine applications put different demands on the installation, such as intake filtration for dust and particle removal, and coatings for corrosion resistance.

Industrial applications require an installation, many times larger than the engine itself

Thrust

Once an aero engine has produced thrust, it can be manipulated in various ways. Thrust reversers are routinely used to assist deceleration on landing, while reheat and deflection for V/STOL represent more exotic forms of thrust manipulation currently only deployed on military aircraft.

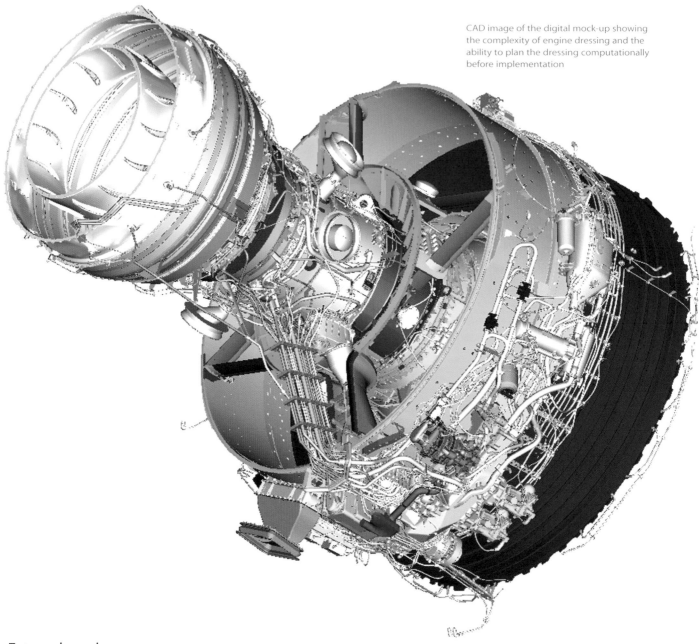

CAD image of the digital mock-up showing the complexity of engine dressing and the ability to plan the dressing computationally before implementation

Externals and the engine build unit

Engine externals are all the elements on a fully dressed engine that connect the engine accessories and controls:

❭ fuel, oil, and pneumatic pipes

❭ brackets and attachments

❭ wiring looms and attachments.

The placement and routing of the engine externals is defined using a digital mock-up, which provides both static clash detection to assist positioning the externals and also dynamic clash detection to help demonstrate maintainability.

The engine build unit comprises the dressed engine with externals along with all the interfaces that need to be connected between the dressed engine and the airframe or nacelle:

❭ cabin air ducts

❭ engine mounts and struts

❭ electrical and hydraulic feeds.

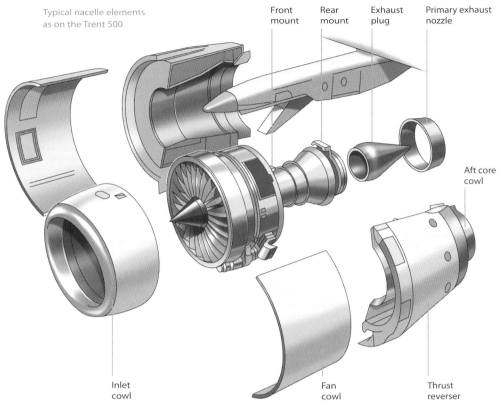

Typical nacelle elements
as on the Trent 500

Front mount
Rear mount
Exhaust plug
Primary exhaust nozzle

Aft core cowl

Inlet cowl

Fan cowl

Thrust reverser

Civil nacelles

A nacelle is a streamlined enclosure that fits around a dressed engine and interfaces with the aircraft structure.

The primary objectives of a nacelle are to

> provide low drag, achieved through aerodynamic design of the nacelle itself and its interaction with the fuselage and smooth surfaces

> ensure good engine performance throughout the aircraft flight and ground envelopes

> reduce engine noise with acoustic treatment of nacelle structure

> prevent ice impact damaging fan blades.

The nacelle must also be manufactured cost effectively, and be easy to install and remove. This must be achieved at the lowest possible weight, while remaining durable and repairable in service.

Nacelles are composed of an air intake, fan cowl doors, nozzle and tail cones, and, optionally, a thrust reverser. Typical civil turbofan nacelles are fitted under the wing on a pylon or fitted to the rear fuselage via a stub wing.

Two further nacelle options are the long (mixed exhaust) nacelle – the Trent 700 – and the short (separate core and bypass jets) nacelle – Trent 800 and 500. Long nacelles can give a performance gain for some engines due to the mixing of the exhaust, and also have a greater acoustic treatment area, at the cost of extra weight and drag.

The fuselage-mounted BR710 on the Gulfstream GV

The underwing-mounted Trent 700, showing the long nacelle option for mixing bypass and core jets before exhausting to the atmosphere

Civil pitot intake geometric features

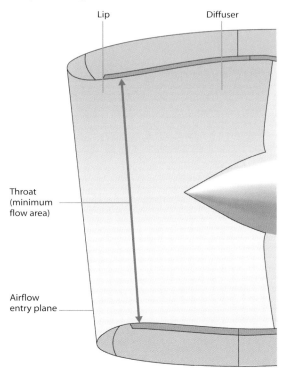

Lip

Diffuser

Throat
(minimum
flow area)

Airflow
entry plane

High incidence climb
Intake turning airflow onto engine axis

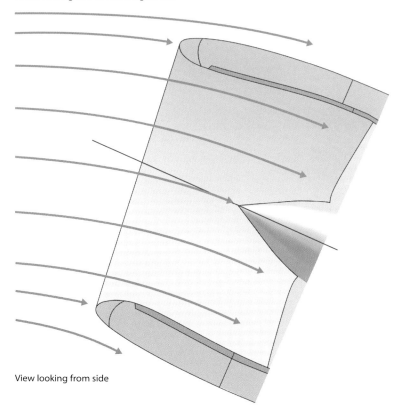

View looking from side

Ground cross-wind operation
Intake turning airflow onto engine axis

View looking from top

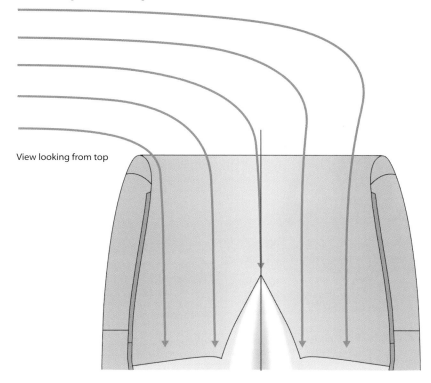

The air intake

The purpose of the intake of civil turbofan engines is to ensure that, under all operating conditions, the engine is supplied with the correct quantity of air, and that the air has sufficient flow uniformity to allow efficient and stable engine operation. The intake design is integrated into the engine nacelle to obtain the lowest level of drag at the operating design point, typically cruise.

For civil turbofans, the optimum intake configuration is a short, near-circular, pitot-type intake. This design is highly efficient for subsonic operation, as low levels of pressure loss are achieved under all operating conditions. Civil pitot intakes are suitable for wing and rear fuselage-mounted nacelles. For tri-jet configurations, s-duct intakes are a design option when the engine is buried in the rear fuselage.

CFD predicted cross-wind condition with flow streamlines

A pitot intake consists of two geometric regions, the 'lip' and the 'diffuser'. The forward section, the lip, is similar in section to an aerofoil and is shaped to guide the airflow into the engine under all operating conditions; the lip is also optimised to prevent flow separation under cross-wind and incidence operation. If flow separation occurs, this produces significant asymmetry in the total pressure within the intake increasing fan blade stresses and, in severe cases, may result in engine surge.

The internal surface of the lip is heated with hot air to ensure that there is no ice accretion, which could shed and damage the engine fan blades (»» 241–243).

The lip contracts to a minimum area, sized for the engine flow requirements, known as the throat. Aft of the throat, the airflow passes into the diffuser where the flow area is increased up to the fan entry plane. The diffuser acts as a settling length to improve the uniformity of the airflow entering the fan. The diffuser section is lined with sound-absorbing, acoustic panels to reduce noise emissions (»» 62).

Civil intake aerodynamic surface shapes are generated as mathematically defined 3D surfaces using CAD tools. The intake surface designs are evaluated and optimised using CFD codes, and the final design is validated by wind tunnel testing.

Before flight testing, to demonstrate that the intake and engine are fully compatible, further testing is conducted over a full range of flow conditions using a machine to simulate high cross-wind speeds.

Intake construction is, typically, an aluminium intake lip for durability and compatibility with the ice protection system, a composite outer skin, an acoustic honeycomb inner barrel, and metal structure bulkheads. The intake is normally bolted onto the engine fan case. Some air intakes might not be circular due to ground clearance constraints and non-uniform accessory distribution around the fan case.

Typical intake construction materials

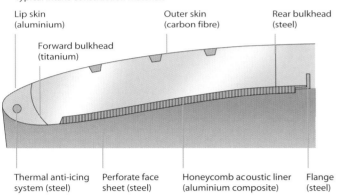

Lip skin
(aluminium)

Forward bulkhead
(titanium)

Outer skin
(carbon fibre)

Rear bulkhead
(steel)

Thermal anti-icing
system (steel)

Perforate face
sheet (steel)

Honeycomb acoustic liner
(aluminium composite)

Flange
(steel)

Following the retirement of Concorde, no immediate replacement existed for civil supersonic air travel. However, any future civil supersonic engines would probably use an external/internal compression intake with variable geometry similar to that used on Concorde and current military aircraft, such an intake provides higher efficiency at supersonic speed.

Fan cowl doors

The fan cowl doors provide a continuous external aerodynamic surface for the nacelle while allowing easy access to all the engine fan case mounted accessories. This access is achieved by having the fan cowl doors hinged to the aircraft pylon.

The top half of each fan cowl door is fire proof as the volume underneath them is a designated fire zone containing the fuel pumps and fuel lines. The airflow under the fan cowl doors must not be obstructed, so that the engine accessories are cooled and ventilated.

Fan cowl doors are typically made from composite materials with a number of access panels for maintenance. Some large fan cowl doors might have powered opening devices; all have 'hold open' rods, which have to ensure safe opening on the ground in winds of up to 110kmh (60 knots).

Thrust reverser

The thrust reverser unit (TRU) has three key functions: to provide a continuous external aerodynamic surface for the nacelle; to provide a fan flowpath for the engine in forward thrust mode; and, of course, to reverse the exhaust flow after the aircraft touches down to assist with aircraft deceleration.

Generally, pilots and airlines want TRUs on jet engines. They can reduce aircraft landing distance, especially on wet and icy runways, while also reducing brake and tyre wear. They improve ground handling on wet and icy runways and taxiways – and improve rejected take-off margins in similar conditions. For military applications, TRUs provide the possibility of operating from bases with shorter runways giving greater operational flexibility.

In civil applications, no certification credit on landing distances is given for TRU fitment; an aircraft landing distance will be determined by the use of anti-lock brakes and aerodynamic

drag devices such as flaps, airbrakes, and parachutes. TRUs are not essential for safe landing; however, they do provide increased safety.

There are four main types of TRU in use today:

> Translating sleeve and pivot door (fan air systems). These are used for large turbofan engines as the majority of the thrust is generated by the fan.

> Target door and pivot door (mixed stream systems). These are used for small, low bypass ratio engines as the split of fan and core thrust is more equal.

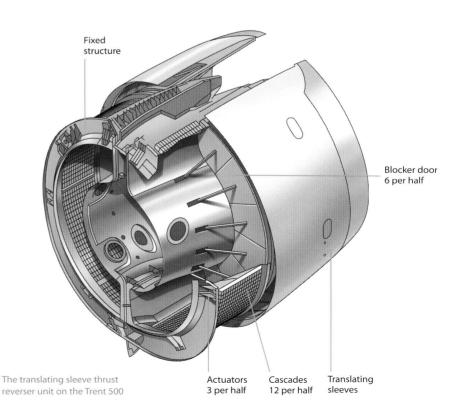

Fixed
structure

Blocker door
6 per half

The translating sleeve thrust
reverser unit on the Trent 500

Actuators
3 per half

Cascades
12 per half

Translating
sleeves

Most large fan engines have 'C' duct TRUs that are split into two halves and hinged to the aircraft pylon, providing access to the engine core components. The TRU also provides a discrete fire zone containing fuel pipes, fuel nozzles, and combustion chambers.

Part of the thrust reverser optimisation process includes ensuring that the hot air/gases neither impinge on the aircraft wing or fuselage nor are re-ingested into the engine intake, which could cause engine surge. It is also important to minimise any lift component from the thrust reverser in order to maximise braking efficiency.

Typically, most TRUs are constructed from carbon composite panels, aluminium structural beams, a metal firewall bulkhead, and a suitable thermal blanket (usually stainless steel) on the inner wall to ensure the epoxy in the carbon composite can withstand the combustion and turbine case temperatures.

Actuation of the translating sleeves or pivot doors is either hydraulic or electrical and three separate locks are provided to ensure there is no TRU deployment in flight. One of these locks will be separately operated, while the other two will be operated and controlled by the engine. For a short nacelle, the TRU also forms the cold, bypass air nozzle.

Nozzles and tail cones

There are two types of nacelle nozzle: the combined cold fan and hot gas nozzle, as seen, for example on the long nacelle of the Trent 700; and the hot gas nozzle, seen, for example, on the short nacelles of the Trent 500 and 800. Tail cones are standard and vary only in their length, cone angle,

Exhaust nozzle and tail cone of a separate jets nacelle

and whether or not they are acoustically treated. A combined nozzle assembly can reduce engine noise emissions by the fitting of acoustic honeycomb panels. Both combined and hot gas nozzles are fitted to the engine LP turbine flange. The tail cone is fitted to the turbine bearing housing at the engine centre and provides a smooth increase in the hot gas exit transitional area.

Military fuselage intakes

Where, as in most military installations, the engine or engines are accommodated within the aircraft, the intakes are incorporated into either the fuselage or wing roots and become a much more integrated part of the aircraft design. As with the civil nacelle intake, the main requirement for such intakes is to supply air to the engine with the minimum loss of pressure and the least increase in aircraft drag. Similarly, for the compressor to operate efficiently and stably, the air delivered to the engine face must be of an acceptable quality in terms of velocity, angularity, and total pressure uniformity. On this form of intake, it is often this last requirement that becomes the most difficult to satisfy because of the physical constraints imposed by the more highly integrated installation and the need for the intake to operate over a wider range of flight speeds and aircraft attitudes.

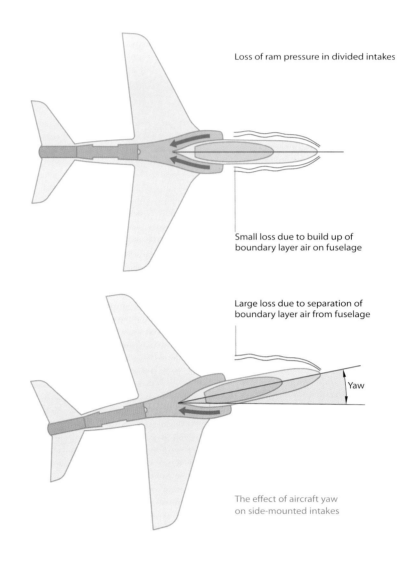

Loss of ram pressure in divided intakes

Small loss due to build up of boundary layer air on fuselage

Large loss due to separation of boundary layer air from fuselage

Yaw

The effect of aircraft yaw on side-mounted intakes

A further, military specific, requirement that is becoming increasingly important is for the intake to conform with the aircraft structure and obscure line-of-sight views of the engine face in such a way as to reduce the aircraft's observability by radar and infra-red detectors. This can lead to highly convoluted intake ducts and unusual intake opening shapes and locations (» 237).

Where the operational flight speed range is mainly subsonic, a pitot intake will normally offer the most efficient solution in terms of pressure recovery and drag. In a single-engined aircraft, this usually involves the use of a divided or bifurcated type of intake set on each side of the fuselage.

One disadvantage of the side-mounted type of intake is that, when the aircraft yaws, a loss of ram pressure occurs on one side of the intake.

This, together with a tendency for the flow over the intake lip to separate when the aircraft is flying at a nose-up incidence, will cause deterioration in both the pressure recovery and uniformity of the air presented to the engine face. The potential influence on engine performance (known as the intake/engine compatibility) demands understanding and attention.

A further example of a fuselage intake where ensuring intake engine compatibility is particularly challenging is the intake for the LiftFan® in the Joint Strike Fighter (F-35 JSF).

The JSF engine installation, showing the two-stage vertical LiftFan®, the two rolls post ducts, and the 3 Bearing Swivel Module

In this installation, the LiftFan is required to operate efficiently and stably behind an extremely short pitot intake at flight speeds of up to 460kmh (250 knots) where the free-stream air is travelling at 90 degrees to the axis of the aircraft. This contrasts with a more normal installation, where conventional, much longer, civil and military intakes operate inside a 55kmh (30 knot) crosswind limitation.

Despite careful intake design, including the use of a rear-mounted intake door to help turn the flow, the non-uniformity of the pressure and the angularity of the air presented

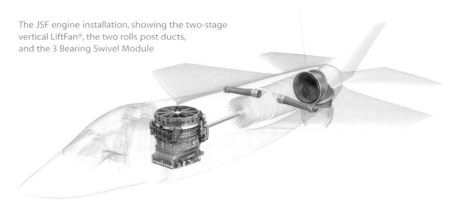

236

to the fan face is more severe than that normally encountered in more conventional intake/engine installations.

To meet this challenge and ensure compatibility between LiftFan® and intake, novel techniques have had to be developed to replicate during ground tests conditions seen in flight.

The pitot type of air intake becomes increasingly unsuitable at higher supersonic speeds due to the severity of the compression shockwave that forms and which progressively reduces the intake pressure recovery. A more suitable type of intake for these higher speeds is known as the external/internal compression intake, which improves the achievable pressure recovery by allowing the air to compress via a series of weaker shockwaves.

As aircraft speed increases still further, so also does the intake compression ratio, and it is necessary, at high Mach numbers, to have an air intake that includes variable geometry features to accommodate and control the changing volumes of air. The airflow velocities encountered in the higher speed range of the aircraft are much greater than the engine can efficiently use; therefore, the air velocity must be decreased between the intake and engine face. The angle of the variable throat area intake automatically varies with aircraft speed, and positions the shockwave to decrease the air velocity at the engine face and maintain maximum pressure recovery. This, coupled with auxiliary air doors used at static and very low speed conditions, allows an efficient match of the intake airflow to the engine demand over the full range of flight conditions.

The under-fuselage-mounted supersonic intakes for the twin EJ200-engined Typhoon

The Panavia Tornado has a twin-engined installation that retains the side mounting but uses an external/internal compression intake combined with a variable-throat area and auxiliary inlets.

The Typhoon, a more agile aircraft, has a twin-engined installation with the intakes mounted under the fuselage. As well as avoiding the problems of fuselage shielding in yaw, this arrangement has the added advantage of using the under-fuselage surface to turn the air into the intake when manoeuvring in a nose-up attitude. This not only off-loads the intake lips to avoid separation at high incidence, but also, at high flight speeds, pre-compresses the inlet air so improving pressure recovery. This intake also has a variable geometry bottom lip, which is used both to improve performance at high incidence and to achieve a better match of the intake capture area to the flight speed – avoiding the need for auxiliary inlets.

Stealth

Enhanced survivability is an important emerging requirement for military aircraft. One way of meeting this requirement is through the reduction of aircraft 'signatures' so the aircraft is less easily detected or traced by potential threats. This type of aircraft is known as a low observable, or 'stealthy', aircraft.

Reducing signature has considerable implications for the engine installation. The major signatures of interest are the radar cross-section (RCS), infra-red emissions – and, to a lesser extent, visibility and noise. The engine air inlet and exhaust duct designs of such stealthy aircraft are driven primarily by the need to achieve the requirements for a minimum signature without too great an impact on aerodynamic performance, weight, and cost. This results in designs that are visibly very different from those of conventional, non-stealthy aircraft.

Cavities, of which the engine air intake and exhaust duct are the largest, are a potentially large source of RCS emissions from a stealthy aircraft. The stealthy engine air inlet duct design obscures the engine fan face using either an inlet entry grill (as on the F-117 Nighthawk), or a convoluted air inlet duct (for example, the F-35 JSF), or blocker vanes immediately upstream of the engine (F/A-18 E/F Super Hornet). These additional surfaces are typically treated with radar absorbent material.

In addition, the engine air intake may itself be located so that it is shielded by the airframe from potential threat radars, and the inlet lips angled to align with the wing leading edges – for example, the X-45 Unmanned Combat Air Vehicle. The inlet duct is also designed to be free of any steps or gaps that may also contribute to the RCS.

Side-mounted intakes for supersonic flow on the twin RB199-engined Tornado

The X-45 UCAV has its intake shielded by the fuselage to reduce detection from the ground[*4]

A Boeing 747 being used as a flying test bed for the Rolls-Royce Trent 800 engine

Similar design features (with the exception of grills) may be used to control the RCS of the engine exhaust system, although the high temperature of the exhaust plume makes this more difficult.

One immediate consequence of the need to geometrically integrate the exhaust nozzle with the trailing edge of the airframe is a trend towards high aspect ratio rectangular nozzles – for example, the F/A-22 Raptor.

The engine exhaust system components and plume, being the hottest parts of the aircraft, dominate its infra-red signature. Consequently, an engine exhaust system designed for a stealthy aircraft is heavily compromised by the need both to shield the view into the hottest parts from the ground and to cool the exhaust plume by mixing it rapidly with the surrounding atmosphere – this will also help reduce jet exhaust noise. In addition, the exhaust system may employ aggressive cooling of the exhaust system components and the application of controlled emissivity materials, which make hot surfaces appear cooler than they actually are.

Flying test beds

A flying test bed (FTB) is usually a production aircraft converted to test a new engine type before the first flight of a new aircraft type. An FTB is fitted with data acquisition equipment and also has a number of simulated systems.

A flying test bed requires a new test pylon or strut adaptor for the specific test engine; it may also need structural modifications.

Historically, before the development of altitude test facilities, a flying test bed was the only means of altitude testing a new engine. There are still specific tests that cannot be done using a ground-based altitude test facility: for example, various nacelle tests and g-load engine tests.

Airframe manufacturers and test pilots normally insist on FTBs for all engine programmes, to evaluate engine operability with representative loads and inlet conditions before the first flight of the prototype aircraft.

Energy and marine installations

Energy and marine engine packages are generally supplied with all engine auxiliaries in place, leaving the builder of the application to provide starter power and fuel, water, and electrical connections.

Intake system

The intake system has to provide protection against snow, rain, and foreign object damage. Marine intakes are corrosion-resistant, often made of composite materials; industrial intakes require dust filters. The intake's large flow area reduces filter pressure loss and avoids ingesting snow or rain.

Cold, humid environments may require heating of the intake to prevent ice formation. Silencing is provided by flow splitters, consisting of sound absorbent material covered in a perforated sheet.

Enclosure

The enclosure provides weather protection (where appropriate), fire protection, and silencing. Ventilation is required to maintain a cooling flow past the engine. Engine accessories are often mounted within the enclosure but off the engine to allow quicker access and maintenance. Access is a key consideration for energy and marine installations to minimise downtime for maintenance (» 262). All enclosures have access panels. Some marine installations, like the WR-21, also allow engine removal via the intake. The design must consider issues such as safe working practices (for example, access control without entrapment), achieving a noise level below 80dB at 1m, and avoiding dangerous hot surfaces.

Although marine and industrial installations are not concerned with the flight speed aspects so important to the aero engine, size remains an issue. The WR-21 enclosure was designed so that it would fit in the footprint of existing marine engines.

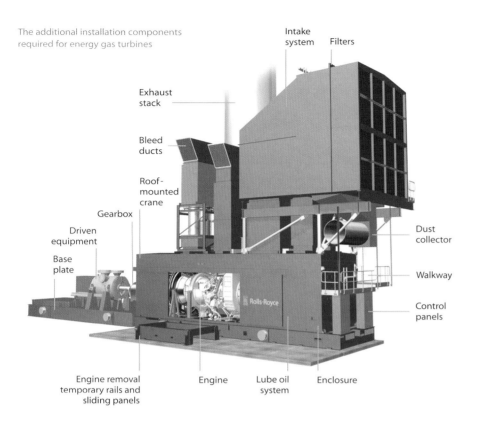

The additional installation components required for energy gas turbines

- Exhaust stack
- Bleed ducts
- Roof-mounted crane
- Gearbox
- Driven equipment
- Base plate
- Intake system
- Filters
- Dust collector
- Walkway
- Control panels
- Engine removal temporary rails and sliding panels
- Engine
- Lube oil system
- Enclosure

Base plate

The base plate, often of steel construction, allows transportation of the package structure. On land, base plates are installed on a concrete base several metres thick to maintain alignment of the drive train and reduce vibration. Offshore, this support is provided by the oil platform or the ship structure.

Exhaust system

The exhaust gases may pass through a heat recovery steam generator. In a combined cycle, a steam turbine generates more electric power; in cogeneration, steam heats a process such as a paper mill. On the WR-21, the exhaust gases are collected and passed to a recuperator that uses the heat energy from the exhaust to pre-heat the combustion air. This improves fuel economy – and, as a side benefit, reduces the exhaust temperature and, therefore, the infra-red signature.

The marine WR-21 installation with recuperation and intercooling

Fire precautions

All gas turbine engines and their associated installation systems incorporate features that minimise the possibility of an engine fire. It is essential, however, that if a fire does occur, it can be detected immediately and rapidly extinguished – and also that there are means of preventing it spreading. For aero engines, the detection and extinguishing systems must add as little weight to the installation as possible.

The main considerations for energy and marine installations are retention of extinguishing fluid, while achieving 'dilution ventilation' for gas fuel leakage. Flame ionisation detectors are used for flame detection.

Prevention of engine fire ignition

Most of the potential sources of flammable fluids are isolated from the 'hot end' of the engine. External fuel and oil system components and their associated pipes are usually located around the fan casings, in a 'cool' zone, and are separated by a fireproof bulkhead from the 'hot zone': the combustion, turbine, and jet pipe areas. Both zones are ventilated to prevent the accumulation of flammable vapours.

All pipes that carry fuel, oil, or hydraulic fluid are made fire-resistant or fireproof to comply with fire regulations, and all electrical components and connections are made explosion-proof. Sparking caused by discharge of static electricity is prevented by bonding all aircraft and engine components – this gives electrical continuity between all the components and makes them incapable of igniting flammable vapour.

The powerplant cowlings are provided with a drainage system to remove flammable fluids from the nacelle or engine bay, and all seal leakages from components are drained overboard so that fluid cannot re-enter the nacelle or engine bay and create a fire hazard.

Spontaneous ignition can be minimised on aircraft flying at high Mach numbers by ducting boundary layer bleed air around the engine. However, if ignition should occur, this high velocity air stream may have to be shut off as it would otherwise increase the flame intensity, and reduce the effectiveness of the extinguishing system by rapidly dispersing the extinguishing fluid.

Cooling and ventilation

The primary function of the ventilation system, which is designed to strict safety and regulatory requirements, is to purge any flammable vapours from the engine compartment. The nacelle or engine bay is cooled and ventilated by air being passed around the engine and then vented overboard. Convection cooling during ground running may be provided by an ejector system.

Fire detection

The rapid detection of a fire is essential to minimise the period before the engine is shut down and the fire extinguished. However, it is also extremely important that a detection system does not give false fire indications as these lead to unnecessary engine shutdown.

A detection system may consist of a number of strategically-located detector units, or be of the continuous element (gas-filled or electrical) sensing type that can be shaped and attached to pre-formed tubes. The sensing element can be routed across outlet orifices, such as a zone extractor ventilation duct.

In the case of electrical systems, the presence of a fire is signalled by a change in the electrical characteristics of the detector circuit, dependent on the type of detector: thermistor, thermocouple, or electrical-continuous element. The change in temperature creates the signal that, through an amplifier, operates the warning indicator.

The gas-filled detector consists of stainless steel tubing filled with gas absorbent material; in the event of a fire or overheat condition, the temperature rise will cause the core of the sensing loop to expel the absorbed active gas into the sealed tube causing a rapid increase in pressure. This build-up of pressure is sensed by the detector alarm switch.

High temperature environments may render thermistor or thermocouple fire detection systems ineffective. Here, thermal detectors that sense either a temperature rise or a rate of temperature rise may prove more suitable. Alternatives to the above types are surveillance detectors that respond to ultra-violet and/or infra-red emissions from a fire.

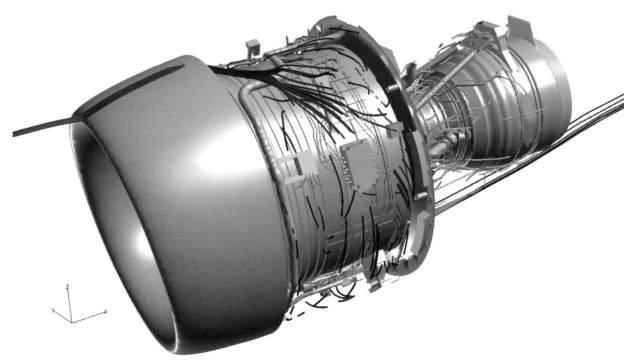

A typical ventilation airflow computer prediction. Advances in computational power, coupled with the extensive use of CAD, allow accurate modelling of these large and complex systems.

Fire containment

An engine fire must be contained within the powerplant and not spread to other parts of the aircraft. The cowlings that surround the engine are usually made of carbon fibre composite. During flight, the airflow around the cowlings provides sufficient cooling to render them fireproof. However, the cowling must be able to contain a fire for a limited period even when the aircraft is stationary on the ground. The nacelle is compartmentalised by fireproof bulkheads, which are designed to prevent the spread of fire.

Fire extinguishing

If a fire is detected, the engine is throttled back to idle. The pilot isolates and shuts down the engine, and the fire extinguisher is operated. The extinguishing fluid is discharged from pressurised containers located outside the fire risk zone through a series of perforated spray pipes or nozzles into all parts of the nacelle. After a fire has been extinguished, the engine remains shut down as any attempt to restart it could re-establish the fire.

Engine overheat detection

Turbine overheat does not constitute a serious fire risk. Detection of an overheat condition, however, is essential to enable the pilot to stop the engine before mechanical or material damage results. A warning system of a similar type to the fire detection system, or thermocouples suitably positioned in the cooling airflow, may be used to detect excessive temperatures. Thermal switches positioned in the engine overboard air vents, such as the cooling air outlets, may also be included to give an additional warning.

Ice protection

Icing of the engine and the leading edges of the intake duct can occur when flying through clouds containing supercooled water droplets or during ground operation in freezing fog. Such ice formation can considerably restrict the airflow through the engine, causing a loss in performance and possible malfunction of the engine. Additionally, damage may result from ice breaking away and being ingested into the engine or hitting the acoustic material lining the intake duct. It is also a threat for energy and marine installations in cold weather.

A typical test on a fireproof silicon seal. Such seals can withstand fifteen minutes application from a large propane burner without losing pressure.

Analyses are carried out to determine whether ice protection is required and, if so, what heat input is required to limit the build-up of ice to acceptable levels

Areas typically considered for ice protection

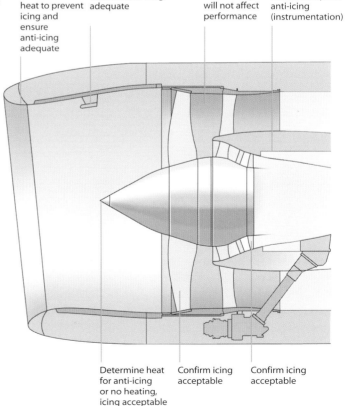

Determine heat to prevent icing and ensure anti-icing adequate

Confirm anti-icing adequate

Ensure icing will not affect performance

Ensure adequate anti-icing (instrumentation)

Determine heat for anti-icing or no heating, icing acceptable

Confirm icing acceptable

Confirm icing acceptable

Turboprop installations on the Hercules C-130J

An ice protection system must prevent ice formation within the operational requirements; it must be reliable, easy to maintain, present no excessive weight penalty, and cause no serious loss in engine performance when in operation.

There are two basic types of ice protection: anti-icing systems that prevent the formation of ice, and de-icing systems that allow ice to form before releasing it. The systems use either hot air or electrical power to heat the components. Hot air systems are usually used on turbojet and turbofan engines; these are typically anti-icing systems. Turboprops, which have less hot air available, often use electrical de-icing systems, or a combination of hot air and electrical systems, with some components anti-iced.

Hot air system

The hot air system provides surface heating of those areas of the engine or powerplant where icing is likely to form. Rotor blades rarely need ice protection because any ice secretions are dispersed by centrifugal action. However, if stators are fitted upstream of the first rotating compressor stage, these may require protection. A rotating nose cone may not need anti-icing if its shape, construction, and rotational characteristics means that likely icing is acceptable. The hot air for the anti-icing system is usually taken from the high-pressure compressor stages. It is ducted through pressure regulating valves to the parts requiring anti-icing.

If the nose cone is anti-iced, its hot air supply may be independent or integral with that of the nose cowl and compressor stators. For an independent system, the nose cone is usually anti-iced by a continuous unregulated supply of hot air via internal ducting from the compressor. Spent air from the nose cone anti-icing system may be exhausted into the compressor intake or vented overboard.

The pressure regulating valves are electrically actuated by manual selection, or automatically by signals from the aircraft ice detection system. The valves prevent excessive pressures being developed in the system, and also act as an economy device at higher engine speeds, when hotter air is available, by limiting the air off-take flow from the compressor – so preventing an excessive loss in performance. The main valve may be locked manually in a pre-selected position before take-off in the event of a valve malfunction prior to replacement.

Electrical system

Turboprops often employ an electrical system, as protection is necessary for the propellers but compressor bleed air supply is limited. The surfaces that require electrical heating are the air intake cowling of the engine, the propeller blades and spinner, and, when applicable, the oil cooler air intake cowling.

Electrical ice protection systems for turbofans have been proposed as part of the move towards 'all electric' aircraft. Such systems may permit

extension of the inlet acoustic treatment around the intake lip, but will need to demonstrate cost and weight benefits before being used in preference to the hot air system.

In electrical protection systems, electrical heating pads, consisting of strip conductors sandwiched between layers of neoprene, or glass cloth impregnated with epoxy resin, are bonded to the outer skin of the cowlings. To protect the pads against rain erosion, they are coated with a special, polyurethane-based paint or covered by a thin metal sheath. When the de-icing system is operating, some of the areas are heated continuously to prevent an ice cap forming on the leading edges and to limit the size of the ice that forms on the other areas that are intermittently heated.

Electrical power is supplied by a generator and, to keep the size and weight of the generator to a minimum, the de-icing electrical loads are cycled between the engine, propeller, and, sometimes, the airframe.

Electric de-icing components on the intake lip of a pitot nacelle

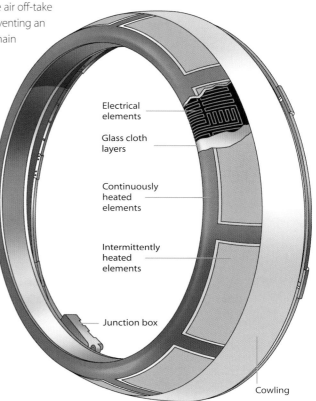

Electrical elements

Glass cloth layers

Continuously heated elements

Intermittently heated elements

Junction box

Cowling

The variable area exhaust nozzle for the reheat system of the RB199

When the ice protection system is in operation, the continuously heated areas prevent any ice forming in those areas, but the intermittently heated areas allow ice to form during their 'heat-off' period. During the 'heat-on' period, adhesion of the ice is broken, and it is then removed by aerodynamic forces.

The cycling time of the intermittently heated elements is arranged to ensure that the engine can accept the amount of ice that collects during the 'heat-off' period and yet ensure that the 'heat-on' period is long enough to give adequate shedding without allowing water to run back and form ice behind the heated areas. A two-speed cycling system is often used to accommodate the propeller and spinner requirements: a 'fast' cycle at high air temperatures when the water concentration is usually greater and a 'slow' cycle in the lower temperature range.

Reheat and variable-area nozzles

Reheat (or afterburning) increases the thrust of an engine to improve the aircraft take-off, climb, and combat performance. This could be achieved by the use of a larger engine, but this would increase the weight, frontal area, and overall fuel consumption; afterburning, therefore, can provide the best method of thrust augmentation for short periods.

The reheat system burns fuel in the volume between the engine turbine system and the exhaust nozzle, using the unburned oxygen in the exhaust gas to support the combustion. The resultant increase in the temperature of the exhaust gas increases the velocity of the jet leaving the propelling nozzle and, thereby, the engine thrust.

The area of the reheated jet pipe and final nozzle is larger than a normal jet pipe and nozzle to accommodate the increased volume of the exhaust gas during reheat. To provide for efficient operation under all conditions, a variable-area nozzle is used. When reheat is selected, the gas temperature increases and the nozzle opens to an exit area suitable for the resultant increase in the volume of the gas stream. This prevents any increase in pressure occurring in the jet pipe that would affect the functioning of the engine.

VTOL, STOL, and vectoring

Vertical take-off and landing (VTOL) or short take-off and landing (STOL) are desirable characteristics for any type of aircraft, provided that the normal flight performance characteristics, including payload and range, are not unreasonably impaired. Until the introduction of the gas turbine engine with its high power-to-weight ratio, the only powered lift system capable of vertical or short take-off and landing (V/STOL) was the helicopter rotor.

Early in 1941, Dr A.A. Griffith envisaged the use of the jet engine as a powered lift system. However, it was not until 1947 that a light-weight jet engine, designed for missile propulsion, existed and had a high enough thrust/weight ratio to be incorporated by Michel Wibault into his ground attack 'Gyroptere' concept. From this early design concept were developed the Pegasus engine and the Harrier fighter aircraft, which after many years of service is only now being superseded by the JSF.

Other uses for thrust vectoring

The need to develop thrust vectoring nozzles has arisen to satisfy other requirements additional to V/STOL capabilities: Providing a conventional take-off and landing (CTOL) aircraft with an enhanced manoeuvring capability for improved combat effectiveness, Supplementing, reducing, or replacing conventional aircraft aerodynamic control surfaces in the interests of reducing weight or drag, or improving aircraft stealth characteristics.

The nature of the design solutions are highly dependent on the needs of the particular application.

Methods of providing powered lift

The Pegasus engine, although the most widely recognised V/STOL concept to enter operational service, represents only one of many ways of providing powered lift:

> swivelling engines

> using bleed air from the engines to increase circulation around the wing and hence increase lift for STOL operations

> using specially designed engines for lift only

> driving a remote lift system, either from the engine or by a separate power unit

> deflecting (or vectoring) the exhaust gases and, therefore, the thrust of the engine.

Among the particular installation challenges posed by STOVL operation using the direct lift principle is the phenomenon of hot gas ingestion. This arises when the hot re-directed exhaust from the engine interacts with the ground, airframe, and external cross-flow in such a way that it is ingested back into either the lift-system or main engine inlet. Hot air ingested in this way reduces the thrust available from the engine and could destabilise the compression system.

Though used predominantly in military applications, future civil aircraft may start to make use of the more exotic forms of thrust manipulation. This would depend on whether short field performance (take-off and landing) becomes sufficiently important to offset the attendant payload and range penalties, and wether the resultant noise signature problems due to thrust manipulation are resolved.

Swivelling engines

The V-22 Osprey's two turboprop propulsion units mechanically swivel through 90 degrees on the wingtips. With the engine nacelles vertical, the aircraft can take off and land vertically, but once airborne, the engine nacelles rotate forward, converting the aircraft to a conventional turboprop capable of twice the forward speed of a helicopter.

Special engines for lift

In the ShinMaywa seaplane, a dedicated gas turbine engine is used to power a ducted fan that drives airflow to the wing and tail control surfaces. This boundary layer control air is used to maintain lift and control surface effectiveness at low aircraft forward speeds. It does this by helping to turn the propeller air stream over the wing, and energise flow over both vertical and horizontal tail control surfaces.

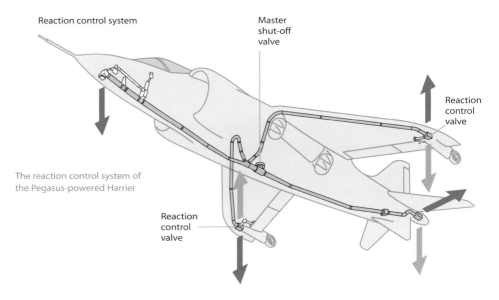

Reaction control system

Master shut-off valve

Reaction control valve

The reaction control system of the Pegasus-powered Harrier

Reaction control valve

Driving a remote lift system

The JSF LiftFan® system transmits 25,000 shaft horsepower from the main engine, via a clutch, to drive a vertically mounted two-stage fan. Pressurised air from the fan is exhausted via a variable area vane box nozzle that can vector between 14 degrees forward and 50 degrees aft in less than two seconds.

The unit produces up to 18,000lb of thrust for use in short take-off operations, and can be throttled to less than 10,000lb during conversion from conventional to STOVL flight. This throttling is achieved by a combination of reduced input shaft speed, variable inlet guide vanes, and nozzle area variation.

Diverting air to deflecting nozzles

The JSF also has roll post ducts: air diverted from behind the fan of the main engine is ducted into the wings where it is turned through 90 degrees to produce lift. By varying the total flow and the port-to-starboard distribution of the air exhausted by the ducts, the system can also be used to control aircraft pitch and roll attitude. In this system, the vertical thrust produced by both ducts is 3,700lb – this can be switched from one side to the other in less than 0.5 seconds. The Harrier uses a related but much smaller system, where the primary function is aircraft control rather than vertical thrust generation.

Deflecting the main engine exhaust

By means of the JSF's 3 Bearing Swivel Module (3BSM), the thrust from the main engine can deflected downwards to provide up to 18,000lb of direct lift. By varying the rotational orientation of the individual sections of the duct, the resultant vertical deflection angle can be progressively varied from 0 to 90 degrees. At any point in this range, further rotation of the duct sections can be used to vector the exhaust sideways for aircraft trim control during vertical manoeuvres.

Combining these ideas in the JSF

In the JSF, the shaft-driven LiftFan® is used to convert some of the power available from the single gas turbine propulsion unit into lift for STOVL operations. The thrust at the front of the aircraft produced by the LiftFan® is balanced both by thrust from under the wings produced by diverting fan air to the roll ducts and also by deflecting main engine

exhaust vertically at the rear using the 3BSM. Once airborne and with conventional lift from the wings, the drive to the LiftFan® is gradually reduced and thrust from the roll-posts and 3BSM are re-directed rearwards.

The JSF represents state-of-the-art technology for vectored thrust and STOVL flight – the result of almost half a century of development.

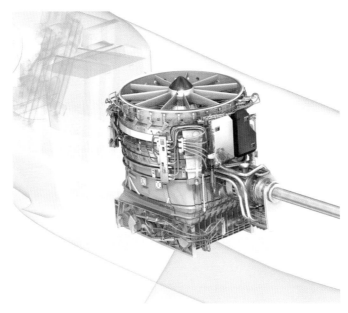

The two-stage, contra-rotating LiftFan® unit provides 18,000lbs of vertical thrust

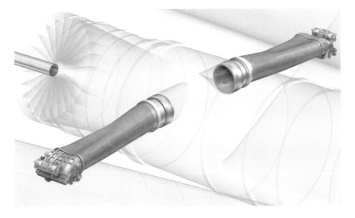

The two roll post ducts provide 3,700lbs of switchable thrust for control of the aircraft in STOVL mode

The 3 Bearing Swivel Module can provide 18,000lbs of vertical thrust in STOVL mode and rotates to give horizontal thrust for conventional flight

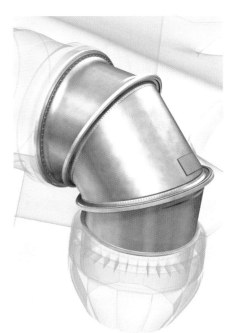

All that has been described so far is no more than a precursor to this point:
an engine ready to produce several decades of useful work.

maintenance

THE DATA CONTAINED IN A COMPLETE SET OF MANUALS FOR A GAS TURBINE CAN AMOUNT TO THE EQUIVALENT OF 250,000 PAGES, CONTAINING SOME 50,000 ILLUSTRATIONS AND 80 MILLION WORDS – ABOUT 100 TIMES LONGER THAN THE COMPLETE WORKS OF SHAKESPEARE. THIS ATTENTION TO DETAIL HELPS ENGINES REMAIN IN SERVICE FOR SEVERAL DECADES.

maintenance

The operations room allows realtime monitoring of engines, diagnosing problems and, where necessary, the ordering of replacement parts to be available when the aircraft lands

Maintenance describes the work required during the engine's service life to ensure it operates safely, reliably, and cost-effectively. Maintenance can be broken into two categories:

〉 line maintenance, which is performed on an installed engine; this is also known as on-wing maintenance for aircraft engines

〉 overhaul, which is undertaken on a removed engine.

Engine management and engine health monitoring are becoming increasingly important and sophisticated aspects of maintenance, helping the operator understand, control, and schedule the work that is necessary on a given engine.

Maintenance of the engine and its systems is carried out according to a comprehensive set of instructions within the maintenance manual. This is based on the manufacturer's recommendations, is constantly checked and updated, and has the relevant certification authorities approval.

During the development of an engine installation, a review of maintenance tasks is undertaken to ensure safe and reliable operation.

A modern engine flies for 13,000 hours between major overhauls – equivalent to seven million miles or flying to the moon and back over 14 times

On-wing maintenance

On-wing maintenance can be divided into two categories: scheduled and unscheduled.

Scheduled maintenance

Scheduled maintenance is a fundamental constituent of the operation of all gas turbines. As part of engine certification, engine manufacturers have to define the minimum standard of scheduled maintenance required to operate the engine. For aerospace applications, this standard is defined using a process of analysis called MSG3 (Maintenance Steering Group 3) – the third evolution of this type of analysis since it was first used in the 1960s. MSG3 divides the engine into all of its systems and subsystems as defined by the Air Transport Association. All the functions of each system are considered along with the possible functional failures, their effects, and causes.

A prime consideration in MSG3 analysis is whether the failure effect is evident or hidden to the crew during normal operating duties. Hidden failures effects are far more likely to generate some form of scheduled maintenance; a maintenance task is mandatory for any hidden failure that has a possible safety impact.

For each function, the failure effects are categorised (for example, safety, operation, or economy) and then any possible maintenance actions reviewed, such as:

> cleaning

> inspection

> functional checks

> lubrication

> restoration

> discard.

If a maintenance action is applicable and cost-effective, the interval at which it needs to be done is calculated. These intervals are calculated from design reliability figures, test data, and previous service experience from similar designs in service, and are specified in cycles, hours, or calendar time.

To aid planning, maintenance tasks are grouped at specific intervals common across the installation or aircraft. Aircraft maintenance intervals are given letters – two common examples are 'A' check (750 hours) and 'C' check (24 months).

The analysis is reviewed at stages by Maintenance Working Groups consisting of the airlines, airworthiness authorities, and the aircraft and engine manufacturers. The working groups make the final decision on maintenance, which is then ratified by an Industry Steering Committee, consisting of all airline operators and the airworthiness authorities.

The list of maintenance tasks is compiled into a Maintenance Review Board Report (MRBR) defining the minimum standard of scheduled maintenance that an airline must accomplish to operate the aircraft. The aircraft manufacturers also produce a maintenance planning document. This document contains all the MRBR tasks and can be customised to suit individual airlines; it forms the basis of the maintenance planning systems that airlines use across the whole aircraft.

Human factors

As with any activity, whenever maintenance is carried out, there is always the potential for human error. One of the objectives of both the engine design and the maintenance programme is to reduce the opportunity for error. Human-centred design considers likely errors, such as the incorrect installation of a component or seal ring, and mitigates the likelihood of these through design. For example, all line replaceable units have integral features that prevent incorrect installation; similarly, adjacent seal rings are sized either to be interchangeable or so that they are significantly different and, therefore, less likely to be installed in the wrong location; also features are included to extend maintenance intervals, so reducing the degree of manual intervention and inspection. In addition, the carefully designed and structured documentation in the maintenance manual minimises the risk of human error.

Multi-engine maintenance, where maintenance tasks have to be carried out on more than one engine of an aircraft at once, carries specific rules and guidelines in the maintenance documentation. Precautions include using different maintenance crews on each engine and independent checking of the tasks to avoid possible errors during maintenance.

Unscheduled maintenance

Unscheduled maintenance is maintenance that was not part of the normal programme. This can be prompted by observed indications from the operators, remote engine health monitoring services, or onboard maintenance indications from the engine's built-in test equipment (BITE).

Unscheduled maintenance can cause delays to operators; therefore, it is important that troubleshooting advice is accurate, concise, timely, and backed up by the necessary logistic and technical support required by the operator. Unscheduled maintenance can range from replacement of a fan blade that has foreign object damage, through to borescope inspection of a compressor following a response to a health-monitored change in HP compressor efficiency.

Increasingly, modern FADEC (» 176–179) and BITE systems give more timely and sophisticated warning of any need for unscheduled maintenance. Paradoxically, this allows airlines to schedule the unscheduled maintenance, planning it into their operation and so avoiding delays to their schedule.

Condition monitoring

Today's modular engines are predominantly managed on-condition: a fixed life for engine removal is not specified – instead, the condition of the engine is monitored and engine removal and overhaul initiated as a result of documented inspection indications (for example, turbine component oxidation) or trend monitoring (for example, turbine gas temperature margin). However, high-energy rotating components, such as discs, do have a specified mandatory life and this must not be exceeded. When this specified life is reached, the engine can be overhauled and only the relevant components replaced or refurbished.

Condition monitoring devices must give indication of any engine deterioration at the earliest possible stage, and enable the area or module in which deterioration is occurring to be identified. This facilitates

quick diagnosis, which can be followed by scheduled monitoring and programmed rectification; the aim is to avoid shutdown, with resultant loss of service, and to minimise secondary damage. Monitoring devices and facilities can be broadly categorised as control room or flight deck indicators, engine performance recorders, and remote indicators.

Flight deck and control room indicators

These are used to monitor engine parameters such as thrust or power, rpm, turbine gas temperature, oil pressure, and vibration. Other devices may be used, for example:

⟩ accelerometers for reliable and precise vibration monitoring

⟩ radiation pyrometers for direct measurement of turbine blade temperature

⟩ return oil temperature indicators

⟩ remote indicators for oil tank content

⟩ engine surge or stall detectors

⟩ rub indicators to sense eccentric running of rotating assemblies

⟩ electronic (in-line) oil system magnetic chip detectors (EMCDs).

In-flight or in-service recorders

Selected engine parameters are recorded during operation. The recordings are processed and analysed for significant trends indicating the commencement of a component or system failure. One such recording device is the time/temperature cycle recorder; this accurately records the time spent operating at critical high turbine gas temperatures, thus providing a more realistic measure of hot-end life than that provided by total engine running hours.

Automatic systems, known as condition monitoring systems, record certain additional pressure, temperature, and flow parameters.

Many of the electronic components used in modern control systems have the ability to monitor their own and associated component operation. Any fault detected is recorded in its built-in memory for subsequent retrieval and rectification by the ground crew. On aircraft that feature electronic engine parameter flight deck displays, certain faults are automatically brought to the flight crew's attention.

Engine condition inspection devices and indicators

Several types of borescope are used for engine internal inspection: they can be flexible or rigid, designed for end or angled viewing, and, in some instances, adaptable for still or video photography. Borescopes are used for examining and assessing the condition of the compressor and turbine assemblies, nozzle guide vanes, and combustors, and can be inserted through access ports located in the engine main casings .

The engine condition indicators include magnetic chip detectors, oil filters, and certain fuel filters. These indicators are used to substantiate indications of failures shown by flight deck or control room monitoring and in-service recordings. For instance, inspection of oil filters and chip detectors can reveal deposits that are early signs of failure. Some maintenance organisations log oil filter and magnetic chip detector histories and catalogue the yield of particles. Similarly, fuel filters may incorporate a silver strip indicator that can be used to detect any abnormal concentration of sulphur in the fuel.

The service data captured by the engine condition monitoring system is also used to assess the health of the engine. Data is corrected to nominal operating conditions and power settings using an engine model. Variation trends can then be used to detect changes due to degradation of internal components. This data can also be used to determine the shaft speed and TET margins, ensuring that the operator recognises when engine margin has reduced to a point where the engine needs to be removed for overhaul.

ETOPS map

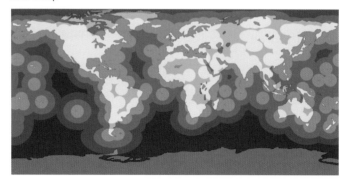

ETOPS limits for 60-, 120-, and 180-minute diversions

Engine health monitoring data is key to successfully managing a large fleet of engines and avoiding operational disruption. Engine health monitoring has several main objectives:

> improve service reliability by reducing in-flight shutdowns, aborted take-offs, and unscheduled engine removals

> drive down the cost of operation by extending component lives

> improve engine workscope management

> provide better customer support.

ETOPS and LROPS

Although long distance operations by twin-engined aircraft are not a recent phenomenon (an early example being Alcock and Brown's pioneering transatlantic flight in a Vickers Vimy, powered by two Rolls-Royce Eagle engines), the early years of commercial transport were dominated by three- or four-engined aircraft. The piston engines of that time were unreliable, and the risk of engine loss during a flight was high.

With the advent of jet-powered transportation in the 1950s, the FAA (Federal Aviation Administration) introduced the '60-minutes rule' for two- and three-engined aircraft. This required that the flight path of these airplanes should should never be more than 60 minutes' flying time from any suitable diversion airport. Inevitably, this resulted in inefficient flight routing.

As the reliability and efficiency of jet engines improved, the risk of engine loss during a flight decreased significantly. This led to a call from operators for a relaxation in the rules, and in the mid 1960s, three-engined jet aircraft were exempted from the 60-minutes rule.

Outside the USA, the ICAO (International Civil Aviation Organization) adopted a 90-minutes limit for twin-engined aircraft, and Airbus developed the twin-engined, wide-body A300. This aircraft proved popular with airlines – maintaining two engines is clearly preferable to maintaining three or four; other twin-engine, long-range aircraft followed, including the Boeing 757, 767, and 777, and the Airbus A310 and A330.

These developments in aircraft design demonstrated to the FAA and ICAO that it is safe for a properly designed twin-engined airliner to fly intercontinental, transoceanic routes. As a result, the FAA introduced ETOPS (extended twin-engine operations) regulations in 1985, setting the conditions that needed to be fulfilled before the grant of a diversion period of 120 minutes – sufficient for direct transatlantic flights. Other airworthiness authorities introduced comparable regulations.

In the late 1980s, the FAA amended the ETOPS regulations to allow a 180-minute diversion period, subject to some technical and operational qualifications. This was adopted by aviation regulatory bodies worldwide, opening 95 per cent of the globe to ETOPS flights.

While the ETOPS regulations were created to apply to twin-engined aircraft, the increased levels of safety and reliability engendered by the ETOPS process are also desirable for three- and four-engined aircraft. As a result, a similar process is being proposed to cover two-, three-, and four-engine aircraft called LROPS (long-range operations).

ETOPS and LROPS requirements

The purpose of these rules is to minimise the risk of an aircraft losing power while in flight. This clearly depends on many factors, including:

> aircraft reliability

> maintenance plans

> distance from aircraft's route to a suitable airport

> crew training

> engine reliability

> hardware standards.

ETOPS approval is given to operators for certain airframe/route combinations. As it imposes a certain financial burden on maintenance and planning, not every operator desires or gains ETOPS approval.

To achieve the engine part of ETOPS approval, an engine manufacturer has to demonstrate the engine's suitability. This will normally involve the demonstration of excellent reliability in service, with evidence from over 250,000 hours of operation. A suitably low in-flight shutdown rate is required, typically less than 0.02 events per 1,000 hours of engine operation for a 180-minute ETOPS rating.

It is also common for minimum engine hardware standards to be defined for ETOPS engines. If experience has shown a certain hardware modification must be incorporated to minimise the risk of engine problems, then that modification may be made mandatory for ETOPS operation.

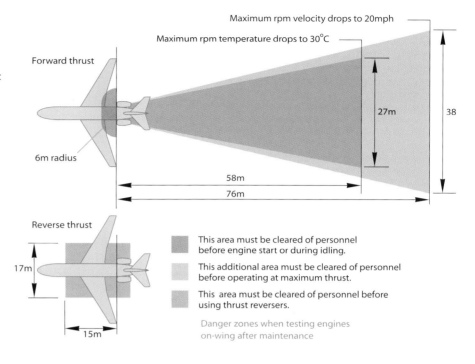

Maximum rpm velocity drops to 20mph

Maximum rpm temperature drops to 30°C

Forward thrust

6m radius

27m

38

58m

76m

Reverse thrust

17m

15m

■ This area must be cleared of personnel before engine start or during idling.

■ This additional area must be cleared of personnel before operating at maximum thrust.

■ This area must be cleared of personnel before using thrust reversers.

Danger zones when testing engines on-wing after maintenance

Early ETOPS

Limiting the use of an engine until it has 250,000 hours of service would be unpopular with airlines, and some engine types have achieved 'Early ETOPS' by demonstrating the reliability of a prototype engine. Typically, this would include engine cyclic tests and a demonstration of problems encountered prior to entry into service – and how those problems were addressed for service.

On-wing engine testing after maintenance

This is undertaken to confirm performance and mechanical integrity and to check a fault or prove a rectification during trouble-shooting. Testing is essential after engine installation, but scheduled testing is not normally required where satisfactory operation on last use is considered the authority for acceptance for subsequent use. In some aerospace applications, this is backed up by specific checks made in cruise or on approach and by evidence from flight deck indicators and recordings.

For both noise and economic reasons, ground testing is kept to a minimum and usually only carried out after engine installations, for trouble-shooting, or to test a system. Improved maintenance methods and engine control system self-test functions, which simulate running conditions during the checking of a static engine, are making the need for engine testing, particularly at high power, virtually unnecessary.

Off-wing overhaul

The purpose of overhaul is to restore an engine so that it meets its performance and reliability requirements. This may result in different levels of refurbishment; the engine is dismantled and parts inspected to determine the need for repair or replacement.

The cost of maintaining an engine in service is an important consideration right from the

initial design stage of a programme, and engine assembly is designed so that overhaul can be completed quickly and cost effectively – a major benefit of the modular engine.

Modular construction and associated tooling enable the engine to be disassembled into a number of modules, or major assemblies. Modules that contain life-limited parts can be replaced with an equivalent, complete module and the engine returned to service with minimum delay. The removed modules are disassembled into mini-modules for replacement of life-limited parts, repair, or complete overhaul as required.

In operation, the engine is managed by an inspection schedule based on manufacturer's recommendations agreed with the air-worthiness authorities and documented in the engine maintenance manual. The engine is removed if its condition is found to be outside set limits, or if engine health monitoring has highlighted that an engine parameter such

as TET, shaft speeds, or vibration has an unacceptable margin. Operators may also choose to remove engines ahead of time in order to achieve 'fleet stagger', so smoothing their engine removal schedule to aid overhaul facility loading.

Stage length, climatic, and environmental conditions all have an effect on the length of time between overhaul, which varies considerably between engine types and operators. When a new engine type or operator enters service, sampling may be conducted to determine the optimum overhaul life. In addition to scheduled overhauls, there are removals that arise from damage during operation.

Cleaning

This is an essential stage in the overhaul and repair of gas turbines. It prepares the parts for inspection and subsequent repair and often includes the removal of coatings, originally applied to protect the parts during

service, that have become damaged or worn. It is also used to improve engine performance without removing the engine. For example, washing the compressor section with suitable detergents and equipment removes dirt and debris from the aerofoil surfaces and so restores efficient airflow through the engine.

Dirt, debris, corrosion, carbon, and oxidation caused by operation of the engine accumulate on engine components during service. Removing these deposits involves the use of a variety of materials ranging from mild detergents and organic solvents to highly active acidic and alkaline chemicals. A high degree of cleanliness is required both to facilitate inspection and to ensure the integrity of a new replacement coating when it is applied.

The choice of cleaning technique is tailored to the surface condition of the parts, the base alloy and a consideration of the environmental impact of that technique – environmental concerns have resulted in the re-formulation or replacement of many processes in recent years. For example, chlorinated solvents that were once widely used have been virtually eliminated as cleaning agents. Other environmentally friendly processes have been added to the cleaning inventory such as dry ice and organic media blasting.

Inspection

The inspection of parts in service, before and after they have been repaired, is critical to the maintenance of engine integrity. While visual, binocular, fluorescent penetrant, and magnetic particle inspection are commonly used, the growth of non-destructive testing (NDT) techniques in recent years means that an increasing variety of inspections can be accomplished both in situ and with the component broken down to piece part level. Borescope, acoustic, ultrasonic, eddy current, x-ray, and holographic inspection techniques all add to the ability to detect flaws and geometric non-conformance. In some cases, surface preparation such as etching is required before inspection.

Engine health monitoring and a generally more sophisticated approach to maintenance mean that modern engines have more in-service inspections than earlier engines. Inspections that traditionally waited for a shop visit now take place during other routine airframe maintenance. The existing borescope access portholes are the obvious point of entry into the engine and the inspection technique is not dissimilar to keyhole surgery.

Demand is growing for this type of restricted access inspection, which necessitates access through a narrow port, about nine millimetres in diameter. The services usually comprise a visual aid borescope, ultrasonic or eddy current probe and couplant delivery system, wires to provide electrical drive for the probe crystal or ferrite core, and a fail-safe wire for retrieval of any part of the probe that may become detached inside the engine.

Borescope viewing resolution has made rapid progress due to CCD (charge-coupled device) chip technology. A chip positioned at the end of the probe allows an electrical signal to be transmitted instead of the traditional fibre optic light. The resultant image is much sharper, enabling more delicate positioning of the probes. The latest borescopes are flexible in that the end can be directed by use of a joystick. The display screen is integral with the kit allowing a single operator to view and manipulate the probe.

Engine removal, whether due to suspected internal damage or because of maintenance schedules, involves high costs for the operators. There is an obvious advantage in allowing an engine to remain in service until defects are revealed by one or more of the following:

- performance analysis
- oil analysis
- borescope inspection
- monitoring of allowable damage.

The future will see a move to telemetry using probes embedded in hardware during manufacture. This technology allows information from the most inaccessible parts of the engine to be monitored, checking for significant change over time. There may be a move from in-service inspections to in-operation inspections – with the possibility of a problem being diagnosed while the aircraft is in flight and the spare part ordered and delivered before the aircraft lands.

Borescope inspection requirements

There are three types of inspection used for borescope inspection: scheduled, special, and unscheduled.

Scheduled inspections

Regular inspections are carried out as part of an approved maintenance schedule, the frequency of which is dependent upon either engine cycles or flight times.

The combustor and turbines are of concern due to the high stresses and temperatures in these areas. All defects should be recorded, ideally on a specific chart, to record any deterioration; if deterioration is noted, assessments are made to establish what action should be taken:

- the engine continues in service to the next scheduled inspection with increased frequency of checks
- the engine is removed within a specified time
- the engine is removed immediately.

Special inspections

Defects may come to light either through service experience or by shop inspection. Instigating special inspections allows these particular defects to be monitored while the engine remains in service.

Unscheduled inspections

Borescope inspection is used to great effect to assess the serviceability of an engine after incidents such as the ingestion of foreign objects, engine surge, or when limits for exhaust gas temperature or rpm have been exceeded.

Repair

Gas turbine components can be subject to wear, impact, handling damage, corrosion, or cracking. A wide variety of techniques is

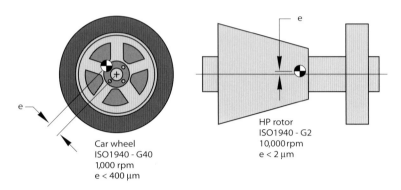

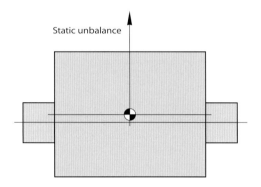

Car wheel
ISO1940 - G40
1,000 rpm
e < 400 μm

HP rotor
ISO1940 - G2
10,000 rpm
e < 2 μm

Static unbalance

Rotor system balancing is a complex but essential task to ensure engine integrity

Three types of unbalance:
static, couple, and dynamic

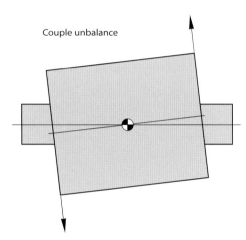

Couple unbalance

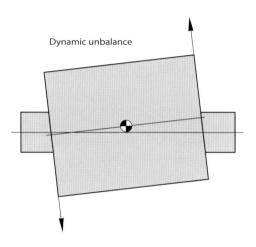

Dynamic unbalance

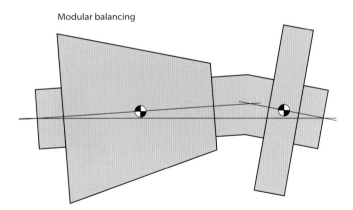

Modular balancing

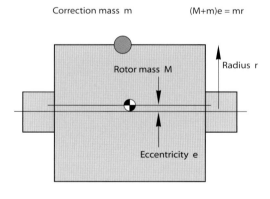

Correction mass m   (M+m)e = mr

Rotor mass M

Radius r

Eccentricity e

Balance correction and module balancing

used to repair engine parts so that they are suitable for further service, thereby avoiding the cost of replacement parts. Some repairs can be carried out on wing, negating the need for engine removal and overhaul. For example, damage to compressor blade tips can be blended using special equipment that uses the borescope ports to gain access.

To restore components to original dimensions, various build-up techniques can be used. These include welding, brazing, metal spraying, and electroplating. The process used depends on the materials of the component and the amount of material to be deposited.

In light-alloy castings, inserts or epoxy fillers may be used depending on the location and type of damage.

Many coatings require reapplication and paints and diffusion coatings are replaced for corrosion protection. Inside the compressor, worn abradable linings are removed and replaced using plasma spray. On large fabrications, cracks can be weld repaired or new flanges or patches welded in.

In some cases, new processes have been developed for repair. Wide gap brazing is used for repair of cracks in turbine nozzle guide vanes that cannot be weld repaired. Micro plasma welding or laser cladding is used for compressor blade tip restoration and seal fin repairs.

Following some weld repairs, heat treatment of the components is necessary either to restore the strength of the material, or temper the weld, or reduce the residual stresses. After repair, it is not always possible to place the whole component in a furnace as this may affect fine limit dimensions or damage coatings; in these cases, local heat treatment with thermal blankets can be used.

Many repairs are affected by the machining of diameters or faces to undersize dimensions or bores to oversize dimensions; the components are then fitted with shims or liners or sprayed with metal coatings of a wear resistant material – after which, the affected surfaces are restored to their original dimensions by machining or grinding.

The increased use of composite materials in aero-engine design, particularly for expensive large structures, has made the specialised field of composite repair increasingly important. The ability to apply disbond, delamination, and patch repairs is necessary and, in order to achieve the correct curing conditions, sometimes requires the application of either pressure through mechanical and vacuum tooling or heat using heat lamps or autoclaves.

Balancing

After overhaul work is done where the engine is disassembled, the main rotating assemblies are rebalanced even if no new parts are installed. Any unbalance in the rotating components is capable of producing vibration and stresses, which increase as the square of the rotational speed.

Any object that rotates will want to spin about its centre of gravity and principal inertia axis; if this is different from the axis as defined by the bearings, vibration will occur. Balancing is the process by which these differences are corrected.

The bearing axis and wheel axis of a car wheel is aligned to less than 400 microns (one micron equals one millionth of a metre and the thickness of a human hair is about 80 microns); in comparison, a typical HP aero-engine rotor is balanced to about two microns.

Much of the effort that goes into designing and balancing the engine compressors and turbines is to ensure that the correct level of balance can be achieved and that the engine maintains this level of balance at all operating temperatures and speeds.

Two terms commonly used in balancing are static unbalance and couple unbalance. Static unbalance occurs when the rotor's centre of gravity is offset from the axis as defined by the bearings. A pure couple unbalance exists when the principal inertia axis is tilted relative to the bearing axis, but the centre of gravity is exactly on the bearing axis. Couple unbalance cannot be detected statically, but at speed it causes a wobbling motion. The combination of static and couple unbalance is often referred to as dynamic unbalance.

Compressors and turbine rotors are made up of a number of components. It is common

for the individual components that make up a rotor to be balanced before being assembled into the complete rotor. This is to minimise, first, the amount of correction required for the complete rotor, and, second, the distributed unbalance in the rotor. The latter is especially important for a rotor that is considered to be flexible at operating speed – such rotors change shape or bend slightly at operating speed and thereby introduce further unbalance.

There are many ways to balance rotors. At component level, metal is often removed by machining. On an assembled compressor or turbine, blade weight differences are used to balance the rotor. In most cases, the mass of the blade is measured and the blades are distributed accordingly. On large fan assemblies, it is common to moment-weigh the blades in three dimensions in order to define the mass and centre of gravity of each blade so that they can be distributed on the disc in a sequence that eliminates both static and couple unbalance. Even with this level of precision, it is common to carry out a fan trim-balance at operating speed. This is because slight variations in blade shape result in variations in blade untwist and lean and, therefore, unbalance at operating speed. Even on the relatively slow moving fan, the aim is to maintain the centre of gravity within four microns of the bearing axis. For final assembly balancing, it is common to add small correction weights to fine-tune the balance of the rotor. Applying a weight at a radius, for example, 20 grams at 100mm radius, results in 2,000gmm. This is the rotor mass, in grams, times eccentricity (mass offset in millimetres), which equals unbalance in gmm.

Because of the modular construction of many modern aero engines, the compressor and turbine are often balanced separately. When the engine is in service, this has the advantage that either the compressor or turbine can be replaced without having to strip the whole rotor. In order to do this, the compressor or turbine is balanced while attached to a dummy rotor that reproduces the bearing span centre of gravity, mass, and principal and diametral moments of inertia of the rotor it replaces. The compressor or turbine rotor assembly is, therefore, corrected

A typical engine running-in profile

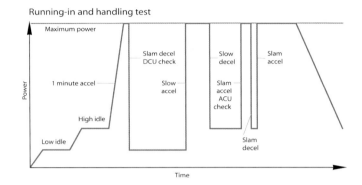

for both its own unbalance and also influence due to geometric errors on any other mating assembly.

New production and overhaul testing
On completion of assembly, every production or overhauled engine must be tested in a ground test cell. The engine is run at ambient temperature and pressure conditions, and the resultant performance figures corrected to International Standard Atmosphere (ISA) sea-level conditions.

This testing is designed to ensure that the engine performance is as expected and does not exceed any engine limits. Tests are also conducted to ensure that engine vibration is acceptable, there are no oil or fuel leaks, and that the engine control system is set up correctly.

Running-in
The running-in handling test is designed to ensure that rotor path linings and other rotating seals are cut in a gradual manner. This consists of progressively faster accelerations and decelerations between idle and maximum power with engine stabilisation times carefully monitored to ensure progressive cutting of seals. This helps performance retention by avoiding damaging rubs during the early service life of the engine.

Accel-decel checks
Engine acceleration and deceleration times between idle and maximum power are recorded to ensure that the engine response is within limits and can meet certification and customer requirements.

Performance tests
Every new production and overhaul engine is put through a pass-off performance test. Typically, this consists of stabilising the engine at six different power levels covering the range between mid power and max take-off. The data gathered at these conditions is processed in near real time by the test bed analysis program, which corrects the data to standard atmospheric conditions, calculates parameters that are not measured directly, like airflow and turbine entry pressures and temperatures, and also derives component efficiencies and pressure ratios. Key parameters are compared against pass-off limits and investigation is carried out if a parameter is outside of one of the three main types of limits:

› airworthiness limits – mandatory limits on shaft speeds and turbine gas temperature

› contractual limits – limits agreed with the aircraft manufacturer, which are stricter than the airworthiness limit to ensure adequate margin in service

› data checking limits – set on a wide number of parameters to ensure that any anomaly in engine data is investigated further.

Trend monitoring of data
The data gathered during the production pass-off testing is used to monitor key engine parameters such as vibration, oil system parameters, and engine performance parameters. There is one data point for every new engine with a ten-engine rolling average line to help identify trends.

The trend monitoring data is used to give advance warning that a series of engines are on a deteriorating trend that may lead to an engine exceeding limits. The data is also used as a check that modifications introduced into an engine or changes in a particular manufacturing process do not give any unexpected results. Unexpected changes in trends require a detailed investigation.

Master test bed calibration

Each engine type has a production test bed designated as the master test bed. This test bed is calibrated so that the engine thrust measured represents the thrust that would be measured on an outdoor test bed. This is done by running a given production engine on both test beds and applying calibration factors.

Customer and reference test bed calibrations

Other test beds can be calibrated against the master test bed. A production test bed calibrated in this way becomes a reference test bed. The same process is used to calibrate customer test beds where typically a lease engine may be used to do testing on both the master and customer test bed.

Monitoring of test bed calibrations

Once calibrated, it is essential that regular checks are made to ensure that a test bed remains within calibration. In addition to maintaining the calibration of test bed instrumentation and test bed configuration control, regular reviews of engine performance trends are essential to ensure that the test bed calibration has not changed significantly.

If a change in the performance trend is discovered, the challenge is to determine whether the root cause is a change in the test bed or a change in the engine itself (due to a revised overhaul procedure, for example). Some changes will be easy to identify – for example, if just one measurement is faulty. Other problems with thrust corrections or engine component changes can be much more difficult to assess and sophisticated data analysis methods have been developed to identify the root cause of problems.

Engine management

Modern gas turbine engines are fitted with condition monitoring facilities to enable the engine to be operated on-condition, allowing it to remain in service providing its condition and performance satisfy maintenance manual acceptance limits. Studies have shown that the most cost-effective method of engine management is to introduce a controlled workscope approach at the engine's shop visit, thereby enabling the engine to return to service for another long, on-condition period. A workscope is a definition and schedule of work for a particular engine taking into account its condition and working environment.

Upon removal of an engine, an appropriate level of workscope is carried out on all modules dependent on the operational life of the engine. Usually, the HP turbine blades will be replaced along with those components that do not have enough residual life to match that of the new blades. To ensure a comparable HP turbine blade life to that of first-run blades, the performance of the engine must also be restored.

A minority of engines are removed before the HP turbine blades reach their thermal life. The decision on the level of workscope on these engines will balance the cost to return them to a serviceable condition and the residual life on the HP turbine blades. There is also an opportunity to swap modules with other engines to optimise the residual life on any particular module.

Engine management programme

The aim of an engine management programme is to define the most cost-effective, in-service maintenance and in-shop work packages to minimise service disruption and maintain optimum levels of reliability and operating cost. Manufacturers continually monitor their engines worldwide so that the experience gained is taken into account when developing maintenance and engine management programmes. The causes of all service disruptions (for example, engine removals, in-flight shutdowns, aborted take-offs, delays, and cancellations) are reviewed and preventative actions initiated. Various methods of solving problems, including special inspections, life limitations, and repair, are considered in conjunction with the development of engineering solutions and then incorporated into the engine manual.

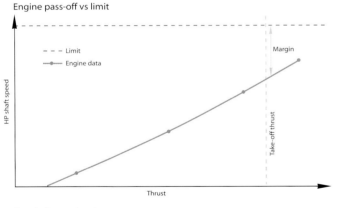

Engine pass-off vs limit

Engine pass-off margin

Trend plot engine data

Trend monitoring plot

On-wing EHM

Much of the information required for engine management is achieved by engine health monitoring while on wing. Collection and analysis of data identifies the engine condition alerting the operator of any potential problem, which can then be rectified, probably still on wing, before the situation becomes critical. The information can be sent to base by satellite while in flight or downloaded as soon as possible after a flight. Up to date information from a large number of sources enables early and accurate diagnosis. The analysis and storage of the data relies on ground-based systems such as COMPASS Navigator™.

Off-wing controlled workscope

Continuously increasing in-service lives requires the development of engine management practices. The workscope is decided by engine history and future requirements.

Weibull analysis is used to establish engine removal drivers and develop the latest engine management practices. This is a method of component statistical failure analysis and distribution. Component failures can be categorised into one of three groups:

> Infantile failures – when there is a risk of failure at low lives with a diminishing risk beyond a certain life. These might include manufacturing errors.

> Random failures – when the risk of a component failing is constant throughout its life. For example, problems caused by foreign object damage.

> Wear-out failures – when there is no risk of failure at low life but a significantly increasing risk of failure at high lives. Thermal fatigue is an example of a wear-out failure.

Engine components can be grouped into these failure categories and a population of the components will exhibit a characteristic failure distribution. It is particularly useful to represent Weibull distributions as a family of straight lines where the slope characteristic is represented by the gradient of the line.

A complex machine such as a gas turbine will contain components with significantly different failure distributions. The reliability of an engine is a function of the interaction of all of its individual components failure distributions. Engine failure distribution can be represented by summing the Weibull lines for all the components. The outcome of this evaluation will maximise in-service life by building in predictability and reliability and by minimising premature removals. A controlled workscope reduces disruption, fuel burn, cost of ownership, and line maintenance, while providing predictable engine removals and spares usage.

Industrial applications

Industrial engines must be robust and able to operate in hostile climates at remote and unmanned stations. Most units are started remotely; therefore, reliability, durability, and availability are of the highest priority.

The maintenance philosophy for industrial engines draws heavily on their aero heritage. The modular concept is retained and there is increasing customer interest in condition monitoring.

In contrast with heavyweight industrial gas turbines (» 135–136), an aero-derivative engine can be removed and replaced quickly – a useful feature as the high-performance gas generator has a necessarily shorter overhaul life than the heavier power turbine and driven equipment. A replacement engine can often be transported by air and road to the site while the installed engine is being removed, thus minimising downtime and spares holding requirements for the customer.

The acoustic package is designed to facilitate rapid engine removal and replacement. Typically, lifting beams and rails are installed to facilitate engine and component replacement, minimising the need for special cranes or lifting gear. Where possible, equipment is positioned to allow easy inspection and maintenance, for example
of duplex filters and oil level – sometimes even while the engine is operating.

Marine applications

Marine applications, like industrial engines, have a need to change engines quickly and easily – the modular approach is again an advantage here; unmanned operation and condition monitoring are also becoming important.

The gas generator and power turbine of the WR-21 consist of 12 interchangeable, pre-balanced modules, which, because of their small size and weight, can be removed via simple routes and new or leased modules fitted in situ, reducing maintenance costs and down time. The engine enclosure is designed for rapid access and permits sideways removal of the gas generator and power turbine. All scheduled maintenance can be performed by the crew and is minimised in line with potential future needs for unmanned engine rooms. Comprehensive borescope facilities allow inspection of all rotating components, the intercooler, and recuperator.

The MT30 is designed for unmanned engine rooms. Condition based maintenance is a feature of the engine design and routine maintenance is limited to checking fluid levels and visual examinations. Internal condition sensors enable the unit to be serviced on an on-condition basis.

The three main groups of Weibull failure distributions can be represented as the labelled 'hazard' functions

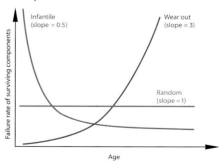

Comparison of Weibull failure distributions

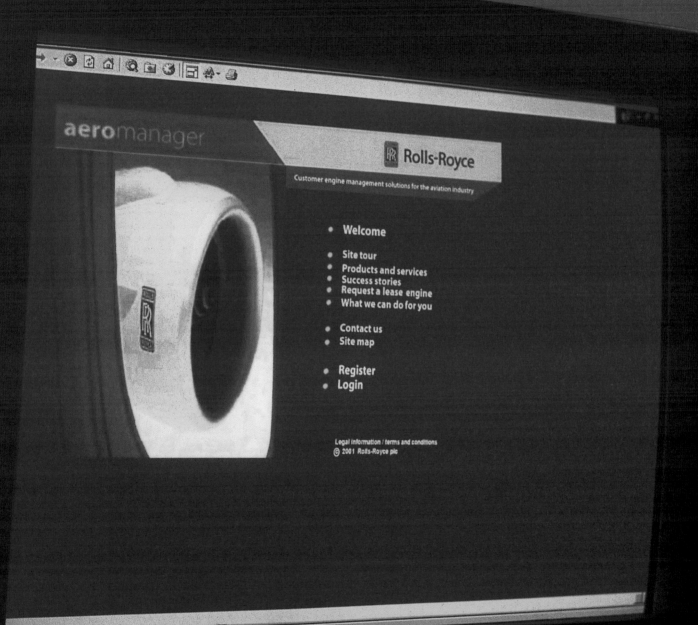

Aeromanager, a web-based portal for service, maintenance, and support documentation

The engines of today are designed to work many years into the future;
the engines of the future are being designed today.

the future

AN ENGINE LAUNCHED TODAY MAY REMAIN IN PRODUCTION FOR MORE THAN TWENTY YEARS. A DESIGN TEAM MUST THEREFORE LOOK TEN, FIFTEEN, TWENTY YEARS INTO THE FUTURE. ON THE OCCASION OF AN ENGINE'S FIRST FLIGHT, THAT ENGINE IS SIMULTANEOUSLY AT THE FOREFRONT OF TECHNOLOGY APPLICATION AND FIVE YEARS BEHIND CURRENT RESEARCH.

the future

A development engineer examines
a test turbine nozzle guide vane assembly

Historically, and for sound engineering and commercial reasons, gas turbine manufacturers have generally adopted an evolutionary approach to engine design. Progress is often incremental; once a technology is proven it is deployed in as many areas as possible. Future development, however, is now likely to involve big changes.

Civil aerospace is at a crossroads, on the brink of a new era where environmental and social factors will take on far greater importance than ever before.

In the 1940s, the Gloster Meteor embodied the technology of the future

The defence sector also enters uncharted territory with the emergence of unmanned aircraft. The marine sector is using gas turbines to drive propulsion systems and vessels very different from the conventional propeller-driven ship.

An artist's impression of a blended wing body aircraft[17]

And energy applications, with their emphasis on efficiency and emissions control, will have a major impact this century.

The propulsion system requirements of civil and military aerospace are moving in quite different directions, although the underlying technologies remain largely common. Advances in materials, 'more electric' technologies, sophisticated design methods, environmentally cleaner and quieter technologies, and the intelligent engine will all influence further developments of the gas turbine. Because of the commonality of the aero-derivative engine, most of these advances will be applicable to energy and marine applications.

The Boeing 787 (top left) and the Airbus A380 (above) will be two of the major civil aircraft of the first quarter of 21st century

The swept fan forces the airflow into the more efficient centre of the blade passage

Today's gas turbines

Due to the time taken to validate new technologies and then incorporate them in to product development cycles, the products entering service now and in the next five years or so, will mostly be developments of those that already exist or are already under development.

Civil aerospace

Compared to previous Trent engines, the Trent 900 incorporates new and significant technologies: the first fully-swept fan design and 3D aerodynamics throughout the compressors and turbines aim to improve efficiency. The swept fan also reduces noise. The HP system contra-rotates relative to the IP and LP systems further improving turbine efficiencies (as previously used in some military and energy applications) and the bypass ratio is increased from previous Trent engines in order to improve specific fuel consumption and further reduce noise. Much of this technology was at the research stage when the Trent 700 and Trent 800 engines were first developed early in the 1990s.

At the smaller end of the market, two-shaft, high bypass ratio turbofans will remain the propulsion system for the majority of narrow-bodied aircraft from small airliners to most business jets, including regional aircraft above about fifty seats. Turboprops will provide power for regional aircraft below this size.

Jet-powered aviation is now an essential part of life, with society completely dependent on the ability to transport both people and goods safely and quickly. The key drivers for civil aviation have always been safety, cost of ownership, and passenger choice. Now, however, environmental factors are becoming increasingly important. Already, the performance and economics of large aircraft are partially compromised in order to meet noise requirements. Future environmental regulations may well impose dramatic changes on the design of aero engines and aircraft. The Advisory Council for Aeronautical Research in Europe (ACARE) has recognised that environmental impact has a tangible cost and has set extremely challenging goals for the aerospace research agenda by 2020, including aggressive environmental, safety, and economic targets. These targets will drive designs of future generations of aero engines and aircraft, thereby accelerating technological progress and forcing adoption of solutions novel compared to today's relatively mature products.

Military aerospace

Like their civil counterparts, the military engines of today and the very near future are for the most part conventional designs embodying proven technologies developed in the late 1980s and early 1990s, but they also include more recent and often highly innovative components.

The EJ200 engine for the Eurofighter Typhoon contains noteworthy technology including all-blisk LP and HP compressors, single-stage shroudless HP and LP turbines, and single crystal turbine blades – as well as brush seals and an airspray combustor derived from the Trent civil engine.

The Rolls-Royce/Turbomeca RTM322 engine for the EH101, NH90, and Apache helicopters also uses state of the art technology: a unique inlet particle separator, which limits foreign object damage and erosion and has no moving parts, a three-stage blisk axial compressor,

The BR710 powers the Gulfstream V (seen here) as well as the Boeing 717 and the BAE Nimrod MRA4

The AE 3007 powers the Embraer RJ135, RJ140, RJ145, and the Cessna Citation X

a compact annular combustor, and a highly efficient gas generator turbine using single crystal alloys.

The V-22 Osprey tilt rotor aircraft fulfils a unique requirement in the military transport arena and required the development of the Rolls-Royce AE 1107C turboshaft engine from the AE 3007. The wing-tip tilting nacelles and self-contained oil system were designed to accommodate vertical and horizontal operation.

Today's defence sector requires a wide range of different aircraft roles from combat to reconnaissance, from helicopters to transports, from tankers to missiles, and from light combat and trainers to the emerging market for unmanned aircraft of all types. The removal of the pilot from vehicles where on-the-spot human interaction is not necessary provides numerous advantages: more stealthy shapes, non-pressurised airframes with less need for safety critical systems, more capacity, longer duration missions, and a significantly reduced training demand. These factors are leading to increased focus on the development of unmanned air vehicles and unmanned combat air vehicles (UAVs and UCAVs).

Engines for energy and marine

Aero-derivative engines for energy applications, such as the Industrial Trent, have their own requirements: for example, very low emissions and the ability to use a variety of fuels. Future industrial engines, although still aero-derivatives, will be very different in some aspects of their design – and indeed are likely to develop technologies that feed back into advanced aerospace products.

Similarly, marine engines are building on the latest aero-engine technologies and using them to develop radical designs of their own. The MT30 and WR-21 are aero-derivatives with significant power and efficiency gains over earlier designs.

Tomorrow's engines

Engines destined to be launched in the medium term are still likely to be largely evolutionary, but based on technologies currently being validated. Cost reduction and performance improvements remain

The EJ200 being assembled at engine build

The RTM322 turboshaft engine powers a range of helicopters including the AugustaWestland Apache seen here

The V-22 Osprey takes off and lands like a helicopter, but can rotate its engine nacelles so that, in flight, it becomes a conventional high-speed, high-altitude turboprop aircraft

important drivers. Emissions are likely to become ever more important for civil aerospace and industrial applications. Looking further ahead, quite revolutionary concepts are being considered in all sectors.

Civil aerospace

The Boeing 787 enters service in 2008, with a new generation of engines such as the Trent 1000. New technology is incorporated into these engines: electric starting, blisk compressors, and significant weight reductions through use of advanced materials and reduced parts count.

Technology validation engines, such as ANTLE (affordable near term low emissions engine), deliver the technology to support the two- and three-shaft architectures and are very much focused on environmental aspects including noise and emissions, thermal, propulsive and component efficiencies, and weight reduction in order to deliver fuel burn improvements and therefore reduced CO_2 production.

The Industrial RB211 has over two million hours of operation and is being used in power stations which require high efficiency and low emissions

The Trent 1000, the fifth generation of the Trent family of three-shaft turbofans

ANTLE – Proving technology

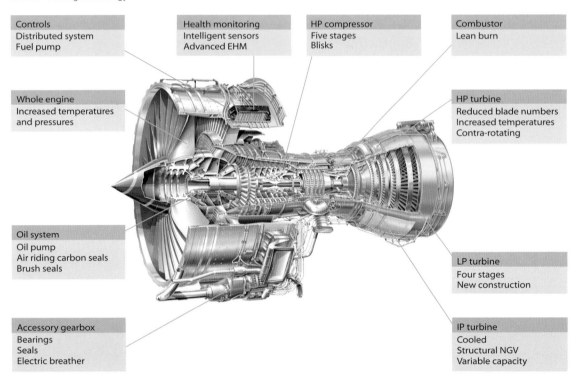

Controls
Distributed system
Fuel pump

Health monitoring
Intelligent sensors
Advanced EHM

HP compressor
Five stages
Blisks

Combustor
Lean burn

Whole engine
Increased temperatures
and pressures

HP turbine
Reduced blade numbers
Increased temperatures
Contra-rotating

Oil system
Oil pump
Air riding carbon seals
Brush seals

LP turbine
Four stages
New construction

Accessory gearbox
Bearings
Seals
Electric breather

IP turbine
Cooled
Structural NGV
Variable capacity

ANTLE takes a low-risk approach to developing key future civil technologies – including a very low-emissions combustor, higher thermal efficiencies for lower fuel consumption, lighter and less complex modules with much lower parts counts, and a distributed control system

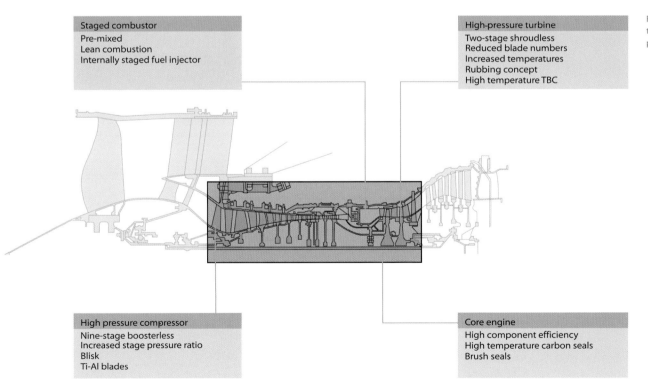

Staged combustor
Pre-mixed
Lean combustion
Internally staged fuel injector

High-pressure turbine
Two-stage shroudless
Reduced blade numbers
Increased temperatures
Rubbing concept
High temperature TBC

Rolls-Royce is a contributor to Germany's E3E research programme

High pressure compressor
Nine-stage boosterless
Increased stage pressure ratio
Blisk
Ti-Al blades

Core engine
High component efficiency
High temperature carbon seals
Brush seals

Everything in front of the HP compressor is based on the Trent 500 engine, but everything behind it – the combustion and turbine systems and advanced controls – is new.

The medium-term will also almost certainly feature new smaller aircraft developments including programmes such as the Brazilian Embraer airframes and the 100-seater aircraft programme in China – although engines will continue to come from the major European and North American suppliers for the foreseeable future.

These types of aircraft will be powered by two-shaft high bypass engines adopting technology developed through UK, US, and German aerospace research programmes such as 'E3E' (Engine – Efficiency, Environment, and Economy), which is fully integrated with the ANTLE programme. This programme features new advanced compressor technology with a strong focus on reducing noise and combustion emissions as well as cost.

However, looking further ahead, in the long-haul market, a very different concept is under consideration: the 'blended wing body', or BWB, aircraft. This offers considerable aerodynamic benefits due to its reduced wetted area and frictional drag. This vehicle could produce the type of step change required to achieve the fuel burn improvements aimed for over the next twenty years. Design constraints, notably the wing depth determined by passenger height, limit the minimum size of a BWB aircraft to above that of conventional wide-body aircraft.

With this configuration, the optimum engine solution may well be quite different from today's large turbofans. In fact, much work has been done in assessing the contra-rotating aft fan.

This BWB and aft fan design improves fuel consumption, weight, and noise. The aft fan configuration lifts the air intake clear of the wing and so enables top-mounted (rather than underslung) engines to be located closer to the fuselage; in this configuration, the wing surface acts as an additional noise shield.

In looking beyond the horizon of the next 25 years, it is necessary to consider alternative fuels both from the viewpoint of emissions and security of supply. Recent studies predict that supplies of kerosene, or a synthetic substitute from natural gas, are sufficient that there will not be a kerosene shortage until 2090. However, environmental pressure may expedite a move towards alternative fuels that are even cleaner than synthetic kerosene but which may themselves bring additional environmental challenges.

Hydrogen and methane are the most obvious alternatives, with methane producing significantly less CO_2 as a combustion by-product than kerosene, and hydrogen producing none at all – though obtaining the hydrogen with traditional energy sources will not eliminate CO_2 production. Using either hydrogen or methane would lead to an increase in the production of water vapour from the aircraft, the effects of which are not yet fully understood. The resulting contrails and their possible

Future wing

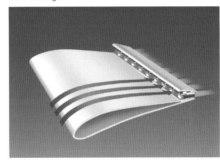

The NASA Mini-engines concept: twenty or more small gas turbines would be positioned along the trailing edge of the aircraft. This would be very efficient aerodynamically.

The TP400-D6 is the most powerful Western turboprop to date, with shaft horsepower in excess of 11,000shp

The Joint Strike Fighter (JSF): a versatile, multi-role aircraft

impact on cirrus cloud formation may also have a detrimental effect on climate change. Together with the practical problems posed in terms of fuel storage, manufacture, and safety issues, commercial use of such alternative fuels is many years away.

Military aerospace

In the medium term, the emphasis will be on versatility in order to contain costs by making one basic aircraft design satisfy several roles. This philosophy has been applied to the Joint Strike Fighter (JSF) multi-role aircraft, the most important fighter programme in the first part of the 21st century, where CTOL (conventional take-off and landing), STOVL (short take-off and vertical landing) and carrier variants aim to provide all military services with versatile and affordable aircraft in large production volumes.

The STOVL variant of the JSF provides its forward vertical lift with the novel Rolls-Royce LiftFan® system, incorporating significant and innovative technology in both its aerodynamic and mechanical design. Aft vertical lift is provided by a three-bearing deflecting nozzle fitted to the main propulsion engine.

The LiftFan® comprises two contra-rotating, high-flow, low-pressure ratio, blisked stages driven by the main propulsion engine and provides around 20,000lbs thrust vertically.

The A400M airlifter is being developed by Airbus Military as a transport aircraft for European military services. The propulsion system requirements for heavy lift capability and short field performance – but with low fuel burn – require four, high-power turboprops of over 11,000hp each. The A400M will use the TP400 turboprop, an engine developed by the Aero Propulsion Alliance (APA) in which Rolls-Royce, Snecma, MTU, Fiat, and ITP are partners.

Looking further ahead, the military market will increasingly split into manned and unmanned vehicles, with growth in the unmanned sector increasing rapidly as the possible uses become proven. The growth in the unmanned sector will cover reconnaissance and combat (both fixed wing and rotorcraft) as well as missiles and space access. The manned sector, however, will predominantly feature growth in reconnaissance and strike.

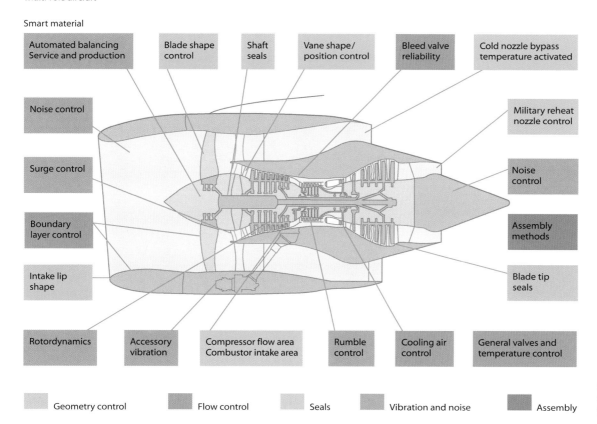

Smart material

Automated balancing Service and production

Blade shape control

Shaft seals

Vane shape/ position control

Bleed valve reliability

Cold nozzle bypass temperature activated

Noise control

Military reheat nozzle control

Surge control

Noise control

Boundary layer control

Assembly methods

Intake lip shape

Blade tip seals

Rotordynamics

Accessory vibration

Compressor flow area Combustor intake area

Rumble control

Cooling air control

General valves and temperature control

Geometry control   Flow control   Seals   Vibration and noise   Assembly

Smart materials and their potential roles in future engines

Engines for energy and marine

The long-term vision for energy is to continue to drive down emissions, including those of CO_2; this requires ever higher efficiency. The ability to burn a wider range of fuels derived from renewable energy sources will also help reduce emissions.

There will be significant developments in marine propulsion, enabling ships to play an increasingly important role in transport, leisure, and defence in an ever more crowded, fast-moving, and networked world. Key areas of technological focus will include further advances in electric technologies. This will lead to high-efficiency, electric systems combined with clean gas turbines, providing very high levels of energy recovery through regeneration.

Current and future technologies

Whatever happens to the jet engine in the future (and, obviously, the longer term that future is, the less predictable it can be), two developments can be regarded as reasonably certain: first, design methods and modelling capabilities will become ever more sophisticated; and second, there will need to be a mix of evolutionary and revolutionary technologies in order to realise new designs and models.

Smart materials

Materials have always been a key factor in improving the performance, reliability, and manufacturability of the jet engine. Smart materials, such as shape memory alloys,

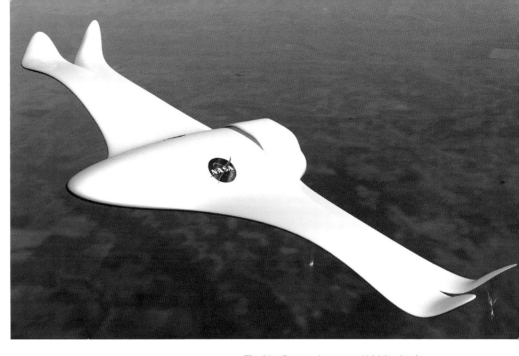

could result in components being able to change shape in response to their environment. This would transform today's approach to engine design, which optimises performance at one operating condition, accepting as a trade-off unoptimised performance at other points in the cycle.

Fans and compressors

Changes in material could also be instrumental in long-term developments of the fan and compressor. Silicon carbide fibre-reinforced titanium could increase both the strength and stiffness of the fan blade so allowing a wider blade chord. This would mean fewer blades were needed, resulting in improved performance and reduced cost.

Titanium is an ideal material in many respects; however, a titanium component rubbing against another titanium component at high temperatures will catch fire. There are,

The 21st Century Aerospace Vehicle, aka the Morphing Airplane: NASA's concept for an aircraft using smart materials and technologies that could change its shape to suit different flight conditions[*4]

therefore, limitations as to where it can be used in the engine. However, research work is currently underway to develop a form of non-burn titanium, which would enable weight-saving titanium blades to replace steel or nickel blades in the rear compressor stages.

The weight of the compressor is also being reduced by the introduction of 'blisks', or bladed discs. Blisks will ultimately be supplanted by 'blings' or bladed rings, which will use advanced materials to provide a seventy per cent weight saving over a conventional design. The bling replaces the bore of the conventional disc with a fibre-reinforced ring.

Blisk and bling technology

Conventional
disk and blades

Blisk - up to 30%
weight saving

Bling - up to 70%
weight saving

Blisks and blings require advanced materials and manufacturing techniques but offer dramatic weight reductions over conventional blade fixings

Combustion

The focus on emissions means that combustion technology will be the object of considerable attention for many years to come. In the medium term, work is being done on evolutionary developments of today's combustion system. Longer term however, if the European aerospace goal of an 80 per cent reduction in the emission of nitrogen oxides (NO_x) from 2001 to 2020 is to be achieved, aero engines will have to learn from the very low emissions systems used in energy applications.

Again in the longer term, the use of ceramics offers the potential for significant temperature increases with minimal cooling. This could result in appreciable efficiency gains and emissions reductions. However, questions over life, strength, fibre capability, and fabrication must first be answered.

Turbines

Turbines have always provided numerous challenges to the designer and remain an area where further developments are critical to improving overall engine performance.

Ceramic combustor

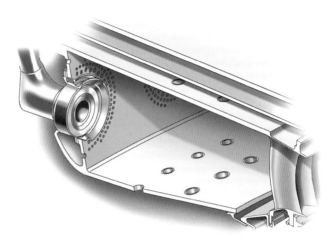

A ceramic combustor could dramatically reduce the required cooling flow

For example, contra-rotating stages can eliminate the need for nozzle guide vanes, reducing weight and part count. There need to be advances in component efficiencies and temperature capability – along with a reduction in cooling air consumption.

As with the fan and combustor, it is possible that ceramics will, sometime in the future, be used in turbines; the benefits would be increased temperature capability and reduced cooling requirements, while the problems to be overcome include fracture toughness and ease of manufacture. HP nozzle guide vanes are likely to provide the initial opportunity for ceramics with the ultimate challenge lying in the realisation of an uncooled HP turbine rotor blade.

Serrated nozzles and scarfed nacelles will help to reduce noise impact around airports

Noise

In the civil context, noise and installation aerodynamics will be particularly important as bypass ratios are increased to reduce exhaust jet velocity and improve fuel consumption. Current and future engines have noise targets that require their performance to be optimised for noise rather than fuel burn in certain points of the cycle. This penalty is often associated with the additional installed drag and weight associated with large diameter nacelles. Avoidance of this penalty requires a different approach to engine installations involving weight reduction using even lighter fan systems and LP turbines, a new approach to thrust reversing, and possibly increased laminar flow nacelles.

In parallel with this approach, noise reduction technologies will play a vital role. One approach is to use a serrated nozzle together with advanced acoustic linings in the intake of the nacelle, further reducing both jet and fan noise respectively. In flight tests, measured results have shown a reduction in fan and jet noise of 4dB and 13dB respectively, albeit with a small performance penalty.

This penalty could be eliminated by further refinement of nozzle serrations to designs that can adapt themselves to the different flight regimes – minimising noise during take-off and maximising efficiency at cruise. To achieve this, the serrations could be stowed at cruise, or could change their shape according to the surrounding air temperature by using shape memory alloys.

More electric engines

Both the civil and defence aerospace sectors are demanding increased levels of electrical power. This is driven by the need for increased functionality and reliability but with reduced weight and cost. This may be achievable by replacing mechanical complexity with elegant electrical solutions. Particular requirements in the civil sector are driven by the demand for increased passenger comfort and facilities, while military aircraft demand increased electrical requirements for network-centric systems, weapons and surveillance equipment, and the growing unmanned sector.

The more electric engine (MEE) follows on directly from the more electric advances of the ANTLE programmes and is expected to deliver step changes in functionality and reliability, while achieving reductions in cost and weight. Reliant upon close engine and airframe integration, these improvements will enable the replacement of traditional modern engine/aircraft systems (that today are individually optimised) with fully optimised electrical systems. An electrically powered environmental control system, for example, is particularly attractive as it provides improvements in fuel burn, while eliminating potential cabin air quality problems caused by supplying cabin air from the engine.

The next step in this evolution at an engine level would be to replace conventional lubrication systems with oil-less, active magnetic bearings (AMBs), ultimately leading to the deletion of the entire oil system and gearbox. A generator, mounted directly on the fan (LP) shaft, would deliver power to the airframe systems and all flight control actuators would also be electric.

However, developments in this field rely heavily on both low weight designs and advancements in electric and magnetic materials, which will be necessary to realise the required temperature capability and reliability. Particular developments in insulation technology, permanent magnet materials, and power electronics are fundamental requirements to achieving the more electric engine and more electric aircraft. These areas are currently being addressed through extensive research and development activity.

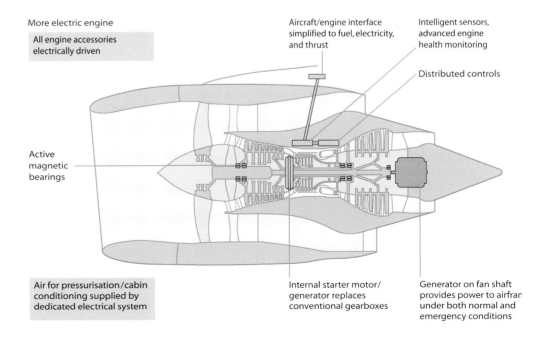

The more electric engine removes mechanical drives and bleed off-takes, creating a more efficient gas turbine

More electric engine

All engine accessories electrically driven

Aircraft/engine interface simplified to fuel, electricity, and thrust

Intelligent sensors, advanced engine health monitoring

Distributed controls

Active magnetic bearings

Air for pressurisation/cabin conditioning supplied by dedicated electrical system

Internal starter motor/ generator replaces conventional gearboxes

Generator on fan shaft provides power to airfram under both normal and emergency conditions

'Whatever is rightly done, however humble, is noble.'

Henry Royce

glossary and conversion factors

a

ACARE – Advisory Council for Aeronautical Research in Europe

AFR – air/fuel ratio

airflow – the movement of air through the gas turbine

b

bling – bladed ring, an advance on the blisk, which replaces the bore of the conventional disc with a fibre-reinforced ring

blisk – bladed disc, a single component comprising both blades and a disc – an advance on the conventional arrangement

c

CAD – computer-aided design

CAM – computer-aided manufacture

CFD – computational fluid dynamics – computer simulation or prediction of fluid flow

combustion gases – the resultant gases after mixing and ignition of fuel and air in the combustor

combustor – the component of the gas turbine in which fuel is mixed with high-pressure air and burnt. Traditionally, 'combustor' has been American terminology and 'combustion chamber' a term used more by British engineers. This book uses both names interchangeably.

compressor – the component of the gas turbine in which the airflow is compressed ready for ignition in the combustor

CTOL – conventional take-off and landing

d

dB – decibel, the logarithmic unit used to measure sound pressure level

e

EGT – exhaust gas temperature

enthalpy – the sum of a system's internal energy and the product of its volume and the pressure exerted on it. Often defined as 'heat content', enthalpy, in fact, includes energy not contained in the system.

EPNdB – effective perceived noise level in decibels

EPR – engine pressure ratio

ESS – engine section stators

f

fan – a large, low-pressure compressor at the front of a turbofan that supplies bypass air to the nozzle

FAR – the inverse of AFR

FEA – finite element analysis, computer analysis of a structure to simulate and predict its behaviour under thermal, static, and dynamic load conditions

Fn and F – Fn is the internationally recognised term for net thrust; this book uses the simpler term, F (see also 'thrust')

g

g – equal to the force of gravity

h

HP – high-pressure

hp – horsepower (see power)

i

ICAO – International Civil Aviation Organization

IP – intermediate-pressure

l

LP – low-pressure

m

Mach 1 – the local speed of sound

n

NGV – nozzle guide vane

NO$_X$ – nitrogen oxides: the term used to describe the sum of nitric oxide (NO), nitric dioxide (NO$_2$), and other oxides of nitrogen that together form a chief component of air pollution

o

OGV – outlet guide vane

on-condition maintenance – maintenance undertaken only when monitoring shows that work is needed

p

power – the ability to do work that results when the gas energy that is used to create thrust in a turbojet or turbofan is instead used to drive a propellor or shaft. Power is measured in horsepower (hp), shaft horsepower, (shp) or kilowatts (kW).

r

referred parameter groups – typically, by ignoring second order effects, the parameters are referred to inlet pressure and temperature; this allows a large number of graphs to be collapsed onto a single plot

rpm – revolutions per minute

s

SO$_X$ – oxides of sulphur including SO$_2$

sfc – specific fuel consumption

shp – shaft horsepower (see power)

stoichiometric ratio – the theoretical air/fuel ratio for perfect combustion, enabling all of the fuel to burn using all of the oxygen in the air

stratosphere – the region of the atmosphere extending from the top of the troposphere up to a height of about 50km

t

TBC – thermal barrier coating

TET – turbine inlet temperature

thrust – one of Newton's principles is that to every action there is an equal and opposite reaction. The expansion of the airflow and combustion gases gas is an action which creates a reaction of equivalent force. This reaction is thrust, which transmitted through the engine to the aircraft. Thrust is measured in pounds force (lbf), kilograms force (kgf), or Newtons (N).

troposphere – the lowest region of the atmosphere, between 8 and 18km above the earth

turbine – the component of a gas turbine that extracts energy from the combustion gases to drive the compressor (and in some cases, a propellor or drive shaft)

turbofan – a development of the turbojet that uses some of the energy from the combustion gases to drive a LP compressor (or fan) that supplies thrust from low-pressure bypass air in addition to the thrust from the high-pressure core system

turbojet – the simplest form of aero gas turbine that relies solely on thrust generated by the core system for propulsion

turboprop – the turboprop uses a propeller to transmit the power it produces – the propeller is driven through a reduction gear by a shaft from a power turbine

turboshaft – the turboshaft uses a power turbine and gearbox to transmit power to a helicopter rotor system or to industrial and marine applications

u

UHC – unburnt hydrocarbon

v

V$_{flight}$ – flight velocity

V$_{jet}$ – jet velocity

VSV – variable stator vane

VIGV – variable inlet guide vane

V/STOL – vertical or short take-off and landing

Unit abbreviations

ft	=	foot
mm	=	millimeter (m x 0.001)
m	=	metre
km	=	kilometre (m x 1,000)
lb	=	pound
g	=	gram
kg	=	kilogram
Btu	=	British thermal unit
K	=	Kelvin
hp	=	horsepower
s	=	second
min	=	minute
h	=	hour
m/s	=	metres per second
W	=	watt
kW	=	kilowatt
MW	=	megawatt
J	=	joule
kJ	=	kilojoule
MJ	=	megajoule
Hz	=	Hertz
N	=	Newton
Pa	=	pascal
kPa	=	kilopascal
MPa	=	megapascal

Conversion factors – exact values are printed in **bold** type

Length	1 in	=	**25.4** mm	
	1 ft	=	**0.3048** m	
	1 mile	=	1.60934 km	
	1 nautical mile (knots)	=	**1.852** km	
Area	1 in²	=	**645.16** mm²	
	1 ft²	=	**92903.04** mm²	
Volume	1 UK fluid ounce	=	28413.1 mm³	
	1 US fluid ounce	=	29573.5 mm³	
	1 Imperial pint	=	568261.0 mm³	
	1 US liquid pint	=	473176.0 mm³	
	1 UK gallon	=	4546090.0 mm³	
	1 US gallon	=	3785410.0 mm³	
	1 in³	=	16387.1 mm³	
	1 ft³	=	0.0283168 m³	
Mass	1 oz (avoir.)	=	28.3495 g	
	1 lb	=	**0.45359237** kg	
	1 UK ton	=	1.01605 tonne	
	1 short ton (2,000 lb)	=	0.907 tonne	
Density	1 lb/in³	=	27679.9 kg/m³	
	1 lb/ft³	=	16.0185 kg/m³	
Velocity	1 in/min	=	0.42333 mm/s	
	1 ft/min	=	**0.00508** m/s	
	1 ft/s	=	**0.3048** m/s	
	1 mile/h	=	1.60934 km/h	
Acceleration	1 ft/s²	=	**0.3048** m/s²	
Mass flow rate	1 lb/h	=	1.25998 x 10⁻⁴ kg/s	
Force	1 lbf	=	4.44822 N	
	1 kgf	=	**9.80665** N	
	1 tonf	=	9964.02 N	
Pressure	1 in Hg (0.0338639 bar)	=	3386.39 Pa	
	1 lbf/in² (0.0689476 bar)	=	6894.76 Pa	
	1 bar	=	**100.0** kPa	
	1 standard atmosphere	=	**101.325** kPa	
Moment (torque)	1 lbf in	=	0.112985 Nm	
	1 lbf ft	=	1.35582 Nm	
Energy/Heat/Work	1 hp h	=	2.68452 MJ	
	1 therm	=	105.506 MJ	
	1 Btu	=	1.05506 kJ	
	1 kWh	=	**3.6** MJ	
Heat flow rate	1 Btu/h	=	0.293071 W	
Power	1 hp (550 ft lbf/s)	=	0.745700 kW	
Kinematic viscosity	1 ft²/s	=	929.03 stokes = 0.092903 m²/s	
Specific enthalpy	1 Btu/ft²	=	37.2589 kJ/m²	
	1 Btu/lb	=	**2.326** kJ/kg	
Plane angle	1 radian (rad)	=	57.2958 degrees	
	1 degree	=	0.0174533 rad	= 1.1111 grade
	1 second	=	4.84814 x 10⁻⁶ rad	= 0.0003 grade
	1 minute	=	2.90888 x 10⁻⁴ rad	= 0.0185 grade
Velocity of rotation	1 revolution/min	=	0.104720 rad/s	

the index

Phototgraphs and illustrations are marked in italics

bibliography, credits, and thanks

Bibliography

Gas Turbine Performance
P P Walsh, P Fletcher, 2004

Jet Propulsion
Nicholas Cumpsty, 2003

The development of jet and turbine aero engines
Bill Gunston, 2002

A history of aerodynamics, and its impact on flying machines
John D Anderson jr, 1997

Gas Turbine Theory
H I H Saravanamuttoo, G F C Rogers, H Cohen, 2001

The design of high-efficiency turbomachinery and gas turbines
D G Wilson and T Korakianitis, 1998

The Magic of a Name, the Rolls-Royce story,
Peter Pugh, 2002

Frank Whittle, invention of the jet
Andrew Nahum, 2004

1904-2004 a century of innovation,
Rolls-Royce, 2004

Credits

All photographs supplied from the archives of Rolls-Royce Group plc except:

1. IAE . . . . V2500 cutaway

2. Turbomeca . . . . RTM322 CF compressor

3. Hamilton Sundstrand . . . . LP CF pump, HP gear pump, circuit board

4. NASA . . . . . . X-45, 21st Century Aerospace Vehicle

5. TWI . . . . Plasma torch

6. Blackwell Science . . . . Performance graphs

7. Embassy Visual Effects . . . . . . Blended wing body

Corbis . . . . Isaac Newton

© BAA plc . . . . Runway

Our thanks

Rolls-Royce Technical Publications have produced five editions of The Jet Engine since 1955, the last published in 1996. In 2004, Nigel Wright had the idea of updating it and Scott Thompson got things moving. Garry Crook and Matt Horlor were the editorial team that technically, creatively, and logistically took the book from idea to reality.

There are many skills that go into a book of this type: Steve Capsey of The Maltings Partnership produced all but a handful of the illustrations; writing and editorial sub-editing came from Jack Nicholas of the Nicholas Allen Strategic Communication Partnership, and book design and typography was by Leslie Jessup, Steve Allen, and Tom Sanderson at Essential Publishing. Repro by Reg French and his team at Essential Repro.

Over 2004 and 2005, over one hundred people at Rolls-Royce worked on this latest edition, a complete rewrite and redesign. People, genuinely too many to mention, contributed their knowledge, hard-won but freely given, often at times when their own work was pressing.

The Jet Engine has been a team effort.
Our thanks to everyone who has made it possible.

Colophon

Many of the illustrations in this book are based on Rolls-Royce CAD images prepared on CADDS or Unigraphics and manipulated in IsoDraw. They were then processed further using Adobe Illustrator and Photoshop – as were the original illustrations.

The typeface is Myriad, the Rolls-Royce typeface.

Photographs are both digital and scanned film.